Guide to the
EPA Refrigerant Handling Certification Exam

Boyce H. Dwiggins

Edward F. Mahoney

Prentice Hall
Upper Saddle River, New Jersey   Columbus, Ohio

Library of Congress Cataloging-in-Publication Data

Dwiggins, Boyce H.
   Guide to the EPA refrigerant handling certification exam / Boyce H. Dwiggins, Edward F. Mahoney.
      p.   c.m.
   ISBN 0-13-011545-2
   1. Refrigeration and refrigerating machinery—Examinations, questions, etc. 2. Technicians in industry—Certification—Study guides.
     I. Title. II. Mahoney, Edward F.
   TP492.3 D.95   2000
   621.5'6'076 21--dc21   99-044801
                                                                                                         CIP

Editor: Ed Francis
Production Editor: Christine M. Buckendahl
Production Coordination: Todd Tedesco, Custom Editorial Productions, Inc.
Design Coordinator: Karrie Converse-Jones
Cover Designer: Allen Bumpus
Cover Art: Allen Bumpus
Production Manager: Patricia A. Tonneman
Marketing Manager: Chris Bracken

This book was set in Palatino by Custom Editorial Productions, Inc., and was printed and bound by R. R. Donnelley & Sons Company. The cover was printed by Phoenix Color Corp.

© 2000 by Prentice-Hall, Inc.
Pearson Education
Upper Saddle River, New Jersey 07458

All rights reserved. No part of this book may be reproduced, in any form or by any means, without permission in writing from the publisher.

Printed in the United States of America

10 9 8 7 6 5 4 3 2 1

ISBN: 0-13-011545-2

Prentice-Hall International (UK) Limited, *London*
Prentice-Hall of Australia Pty. Limited, *Sydney*
Prentice-Hall of Canada, Inc., *Toronto*
Prentice-Hall Hispanoamericana, S. A., *Mexico*
Prentice-Hall of India Private Limited, *New Delhi*
Prentice-Hall of Japan, Inc., *Tokyo*
Prentice-Hall (Singapore) Pte. Ltd., *Singapore*
Editora Prentice-Hall do Brasil, Ltda., *Rio de Janeiro*

CONTENTS

SECTION 1: INTRODUCTION   1
    Who Should Take the EPA Certification Test   2
    Who Writes the Tests?   2
    Type of Test   2
    What You Should Know   3
    Refrigeration System Review   3
    System Review   5
    Evaporator   10
    Be Test Wise—Pass the Test   12
    Temperature-Pressure Chart   13
    Before Test Day   14
    Test-Taking Fear   14
    Relax   15
    Meditate   15
    Distractions   15
    Cramming   16
    Test Yourself   16
    Ground Rules   17
    Discriminate and Eliminate   17
    Change an Answer   19

Test-Taking Techniques   19
Read and Understand   20
Test-Taking Strategies   21
On Test Day   24
Good Luck   25
Health and Safety   25
What Is Ozone?   25
Atmospheric Ozone   26
Absorption of Ultraviolet Radiation   26
Destruction of the Ozone   27
Effects of Ozone Loss on Human Health   28
The Greenhouse Effect   29
The Clean Air Act   29
Stratospheric Ozone Protection—Title VI   30
Significant New Alternatives Policy   31

SECTION 2: CERTIFICATION   33
  CORE Test   34
    General Knowledge   34
    Overview of Issues on the CORE Test   34
    Ozone Depletion   37
    Three Primary Refrigerant Types   38
    Ozone Depletion Potential   38
    The Clean Air Act and the Montreal Protocol   38
    Recovery Devices   39
    Recovery Techniques   40
    Recovery Cylinders   41
    Shipping and Transporting   42
    Dehydration   42
    Leak Repair Requirements   43
    Substitute Refrigerants and Lubricants   43
    Safety   44
    Practice Test   45

  Type I   57
    Small Appliances   57

Contents

    An Overview of the Issues on the Type I Test   57
    Equipment Requirements   59
    Recovery Techniques   60
    Type I Test   61

Type II   74
    High-Pressure Refrigerant Certification   74
    Overview of Issues on the Type II Test   74
    Leak Detection   76
    Recharging   77
    Recovery   77
    Recovery Requirements   77
    Type II Test   78

Type III   91
    Low-Pressure Refrigerant Certification   91
    Overview of Issues on the Type III Test   91
    Leak Detection   93
    Recovery   94
    Recovery Requirements   94
    Recharging   95
    Purge Units   95
    Type III Test   95

SECTION 3: APPENDIX   107
    Section 608 — Refrigerant Recycling Rule   110
    Title VI — Stratospheric Ozone Protection   121
    Sec. 601. Definitions   122
    Sec. 602. Listing of Class I and II Substances (n/a)   123
    Sec. 603. Monitoring and Reporting Requirements   124
    Sec. 604. Phaseout of Production and
            Consumption of Class I Substances   126
    Sec. 605. Phaseout of Production and
            Consumption of Class II Substances   131
    Sec. 606. Accelerated Schedule   133
    Sec. 607. Exchange Authority   134

Sec. 608. National Recycling and Emission
        Reduction Program   135
Sec. 609. Servicing of Motor Vehicle Air Conditioners   137
Sec. 610. Nonessential Products Containing CFCs   140
Sec. 611. Labeling   141
Sec. 612. Safe Alternatives Policy   144
Sec. 613. Federal Procurement   146
Sec. 614. Relationship to Other Laws   146
Sec. 615. Authority of Administrator   147
Sec. 616. Transfers Among Parties to the
        Montreal Protocol   148
Sec. 617. International Cooperation   149
Sec. 618. Miscellaneous Provision   149

SECTION 4: 608 — TECHNICIAN CERTIFICATION PROGRAMS   151

SECTION 5: GLOSSARY   167

SECTION 6: ANSWER KEYS   217
    CORE   217
    Type I   217
    Type II   218
    Type III   218

Section 1

INTRODUCTION

WHO SHOULD TAKE THE EPA CERTIFICATION TEST

The Environmental Protection Agency (EPA) certification test is provided for air-conditioning and refrigeration technicians who work with refrigerants. It is unlawful for any person who is not certified through an EPA-approved test to maintain, service, or dispose of refrigeration equipment.

This certificate also shows your customers that you have the knowledge and skills needed in today's highly technical field of air-conditioning and refrigeration service. Many of the technicians taking the certification examination may not have attended school or taken a test for a number of years. Like any other subject, test-taking can be learned. The two main ingredients to successful test-taking skills are:

Know the subject matter.

Know how to take a test.

For a refresher of the material it is suggested that the following books be used for study reference:

Modern Refrigeration and Air Conditioning–Goodheart Willcox

Refrigeration and Air Conditioning–Prentice Hall

Refrigeration and Air Conditioning Technology–Delmar Publishers

The purpose of this manual is to help you to prepare to take the test. Refer to the Bibliography for suggested reference material.

WHO WRITES THE TESTS?

All the tests are written by qualified individuals with a good knowledge and years of experience in the handling of refrigerants. They are also well versed in the new EPA laws regarding refrigerants. The questions are written by experts in the field, reviewed by other experts, and then pretested by certified technicians.

TYPE OF TEST

All the EPA certification tests are called objective tests. Everyone being tested will be tested with the same type of test consisting of four-part multiple-choice questions. For each question, however, there is only one

Introduction 3

correct answer. The tests are machine scored, which means that the opinion of the proctor or anyone else involved in the test procedures does not influence the outcome.

You will be tested on your knowledge of the procedures and laws associated with refrigerant handling as well as your general knowledge of air-conditioning and refrigeration system theory and repair.

WHAT YOU SHOULD KNOW

Many tests are as much a measure of the *way* you study as your ability to organize a mountain of material. This is especially true of any test that measures your knowledge and mastery of a broad spectrum of material relating to heating, air-conditioning, refrigeration, and ventilation (HARV).

The better you study, the better you will score on a test. There are ways to organize your studying to achieve maximum results in minimal time. There are techniques to use when studying for a test.

Before you can decide how to study for a particular test, it is important to know exactly what you are to be tested on.

The EPA-approved refrigeration certification tests cover everything that you have learned over the years while working as an air-conditioning or refrigeration technician.

REFRIGERATION SYSTEM REVIEW

There are questions on the EPA certification examination about the condition of the refrigerant in various sections of the refrigeration system. There are only six conditions to be considered.

- Low-pressure vapor
- Low-pressure liquid
- Low-pressure vapor and liquid
- High-pressure vapor
- High-pressure liquid
- High-pressure liquid and vapor

The following is a brief overview of these conditions. For the component locations, refer to the callouts (A through F) in Figure 1.

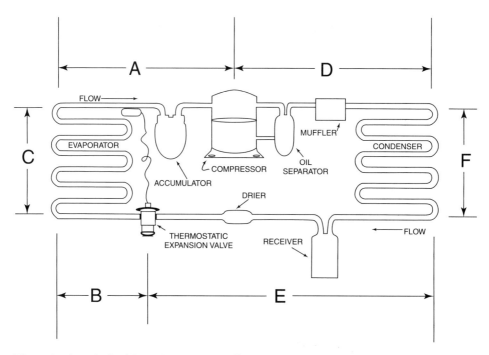

Figure 1 A typical refrigeration system to illustrate component location and condition of refrigerant. (A) Low-pressure vapor (B) Low-pressure liquid (C) Low-pressure liquid and vapor (D) High-pressure vapor (E) High-pressure liquid (F) High-pressure vapor and liquid.

Low-Pressure Vapor

The section of the system from the evaporator outlet to the compressor inlet (A) refrigerant is low-pressure vapor. This includes any auxiliary devices, such as suction line drier, muffler, or accumulator.

Low-Pressure Liquid

The only part of the system that may be considered low-pressure liquid refrigerant is the first row of tubing in the evaporator immediately after the metering device (B).

Low-Pressure Vapor and Liquid

In the evaporator (C) the low-pressure liquid refrigerant picks up heat and is changed to low-pressure vapor.

High-Pressure Vapor

The refrigerant is a high-pressure vapor in the line from the compressor outlet to the condenser inlet (D). This includes any auxiliary devices, such as muffler and oil separator.

High-Pressure Liquid

The high-pressure liquid section extends from the condenser outlet to the metering device inlet (E). This includes any auxiliary devices, such as receiver, drier, and sight glass.

High-Pressure Liquid and Vapor

In the condenser (F) the high-pressure refrigerant vapor, in giving up its heat, is changed to liquid.

SYSTEM REVIEW

As a review, the purpose and function of each component of the refrigeration system as well as the condition of the refrigerant in each will be discussed. For this purpose the compressor will be used as the starting point.

Compressor

The compressor is the component that separates the low side of the system from the high side of the system. The refrigerant enters the compressor as a low-pressure, slightly superheated, vapor. When leaving the compressor, the refrigerant is a high-pressure, highly superheated vapor.

The function of the compressor is to circulate refrigerant throughout the system. The compressor is a pump that is designed to raise the pressure of the refrigerant as well as to circulate it through the system.

According to the laws of physics, when the pressure of a gas or vapor is increased, its temperature is also increased. When pressure and temperature are increased, refrigerant condenses more rapidly in the next component, the condenser.

The four most popular types of compressors are reciprocating, rotary, scroll, and centrifugal.

The Reciprocating Compressor. Each piston in a reciprocating compressor has one suction and one discharge valve. During the intake stroke, also called the suction stroke, the piston draws refrigerant vapor in through the suction valve. At this time the discharge valve is held closed by the higher pressure above it. The suction valve, however, is opened to allow the vapor of low-pressure refrigerants to enter.

When the piston is on the compression, or discharge stroke, refrigerant vapor pressure forces the discharge valve open to purge the refrigerant. During this period, the suction valve is held closed by this same pressure.

Rotary Compressor. The only moving parts in a rotary compressor consist of a steel ring, an eccentric or a cam, and a spring-loaded sliding barrier. The steel ring and cam are housed in a steel cylinder. The steel ring, which has a smaller diameter than the cylinder, is off-center so that one point on the outer circumference is always in contact with the wall of the cylinder. This leaves an open crescent-shaped space on the opposite side between the ring and the cylinder wall.

As the cam rotates, the ring is carried around with it, imparting a rolling motion to the ring on its outer rim around the wall of the cylinder.

The ring does not move very far before uncovering both the suction and discharge ports. A spring-loaded barrier, held snugly against the ring, ensures a seal between the suction and discharge ports.

A port in the cylinder wall is used as an entry for refrigerant vapor from the evaporator to flow into the crescent-shaped space. When the ring is rotated slightly, it will cover up the port trapping the refrigerant vapor in the crescent-shaped space.

A discharge port provides an escape for the entrapped vapor. As the cam continues to rotate and roll the ring around in the cylinder, it pushes the crescent-shaped wedge of refrigerant ahead of it, compressing the vapor as it is forced out through the discharge port to the condenser. At the same time low-pressure refrigerant vapor is entering the suction port.

As the ring continues to rotate, refrigerant vapor is forced out through the discharge port and the cycle is repeated.

Scroll Compressor. The scroll compressor design is a simple compression concept of unique spiral shape scrolls. Two identical scrolls are mated together to form concentric spiral shapes. One scroll is stationary while the other orbits. The orbiting scroll orbits the stationary scroll without rotating.

The scroll orbits in a clockwise (cw) direction to draw refrigerant vapor into the outer crescent-shaped pocket created by the two scrolls. Centrifugal action of the orbiting scroll seals off the flanks of the scrolls.

As orbiting continues, the vapor is forced toward the center of the scroll and the gas pocket becomes smaller and more compressed. When the highly compressed vapor reaches the center, it is discharged into the port in the top of the compressor.

The discharge pressure, forcing down on the top scroll, helps to seal off the upper and lower tips of the scroll. During each orbit of the scroll, several pockets of vapor are compressed, providing a smooth and continuous compression cycle.

The scroll compressor tolerates the return of liquid refrigerant. When liquid enters the compressor, the orbiting scroll separates from the stationary scroll and the liquid is worked toward to the center of the scroll where it is discharged.

Centrifugal Compressor. These compressors produce a pressure through centrifugal effect alone. Compression is caused by whirling the mass of gaseous refrigerant at a high rate of speed, causing it to be thrown outward by centrifugal force into a channel where it is caught. The operation of a centrifugal compressor is similar to the operation of a water pump, but the compressor operates at a much higher speed.

In a centrifugal compressor the molecules of gas are whirling at a high rate of speed and are thrown off the outer edge of a rotating wheel where they are caught in a channel and compressed. A centrifugal compressor consists of a wheel made of several rotors or impellers, each having separate compartments or stages.

Unlike reciprocating and rotary compressors, where the refrigerant vapor molecules are squeezed inside the cylinder by positive action of the piston, compression is produced and maintained by the action of the suction and discharge valves. In the centrifugal compressor, there are no valves and compression is produced by forcing vapor into progressively smaller stages.

Discharge Line

The tube leaving the compressor contains high-pressure refrigerant vapor. Known as the hot-gas discharge line, it connects the outlet of the compressor to the inlet of the condenser.

Oil Separator

Refrigeration oil in the compressor keeps the crankshaft, connecting rods, piston, and other internal compressor parts lubricated. A small amount of oil circulates with the refrigerant through the system. The velocity of the refrigerant carries the oil and returns it to the compressor. In applications where oil circulation through the system is not wanted, an oil separator is found in the compressor discharge line. Its purpose is to trap any oil and return it to the compressor.

Condenser

There are basically two types of condenser, air cooled and water cooled. In either type, heat-laden refrigerant vapor liquefies or condenses in the condenser as it emits heat.

Air-Cooled Condenser. As cooler air passing over the condenser carries its heat away, the vapor condenses. Heat that is removed from the refrigerant in the condenser as it changes from a vapor to a liquid is the same heat that was absorbed in the evaporator as it changed from a liquid to a vapor.

Water-Cooled Condenser. Water-cooled condensers are often preferred because they provide lower condensing pressures and better control of the compressor discharge pressure. Water is generally colder than daytime ambient air temperatures. When cooling-tower supplied, the condensing water can be cooled to a point approaching the ambient wet-bulb temperature.

For a condenser to perform properly, there must be a sufficient amount of cooling medium passing through it, either air or water.

The refrigerant in the condenser is a combination of liquid and vapor under high pressure. The inlet of the condenser must be at the top so the condensing refrigerant will flow to the bottom of the condenser where it is forced, under high pressure, to the liquid line or the receiver.

Liquid Line

The liquid line is the tubing that connects the condenser to the receiver and/or metering device. There may also be a drier, muffler, and/or sight glass found in the liquid line. The refrigerant in this line, as its name implies, is in the liquid state.

Receiver

The receiver, a tanklike storage device, is used in systems that have an expansion valve as a metering device. The receiver holds the proper amount of reserve liquid refrigerant required to ensure proper performance of the

Introduction

air-conditioning system under variable operating conditions. The receiver is located in the high-pressure side of the air-conditioning system between the condenser outlet and the metering device inlet.

Drier

The drier contains a desiccant, a chemical drying agent, that can absorb and hold a small quantity of moisture. It is important that the desiccant is compatible with refrigerants and lubricant used in any particular application.

A screen and/or strainer is included inside the drier to prevent the circulation of any debris throughout the system that may have entered during careless service procedures.

The liquid-line drier contains high-pressure refrigerant in the liquid state. In certain applications a suction-line drier may be used. The state of the refrigerant in the suction line is low-pressure vapor.

Metering Device

The metering device is the part that separates the high-pressure side from the low-pressure side of the system. Its function, as its name implies, is to meter the proper amount of liquid refrigerant into the evaporator.

The most popular metering devices are the automatic expansion valve (AXV), thermostatic expansion valve (TXV), thermal-electric expansion valve (TEXV), and capillary tube (cap tube).

Automatic Expansion Valve. The AXV opens and closes without the aid of an external device. It automatically maintains a nearly constant refrigerant pressure in the evaporator. The opening and closing of the AXV is controlled by refrigerant pressure in the low side of the system. The AXV will not compensate for varying conditions in the low or high side of the system, or for variation in heat loads.

The inlet side of the AXV is a high-pressure liquid refrigerant and the outlet side is low-pressure liquid refrigerant.

Thermostatic Expansion Valve. A popular device for controlling the flow of liquid refrigerant into the evaporator is the TXV. An orifice in the valve, metering the flow of liquid refrigerant into the evaporator, is modulated by a needle-type plunger and seat.

The inlet side of the TXV is a high-pressure liquid refrigerant and the outlet side is low-pressure liquid refrigerant.

Thermal-Electric Expansion Valve. The TEXV is a refrigerant flow control device that is compatible with electronic sensing and monitoring systems. The TEXV is a heat-motor-operated needle valve infinitely positionable in response to an input of 0- to 24-volt signals. An increase opens the valve and a decrease in voltage reduces the flow or closes the valve. The changes in voltage necessary to regulate this valve can be established using various temperature or pressure sensors.

The inlet side of the TEXV is a high-pressure liquid refrigerant and the outlet side is low-pressure liquid refrigerant.

Fixed-Orifice Flow Control Device. A fixed-orifice flow control device can be used to replace other types of flow control devices. Typically the flow control device, such as that manufactured by AeroQuip, is made up of three components: the connector, which attaches to the liquid line; the restrictor, which meters the flow of refrigerant; and the distributor, which provides flow to the evaporator.

The inlet side of the fixed orifice is high-pressure liquid refrigerant and the outlet side is low-pressure liquid refrigerant.

Capillary Tube. The capillary tube, often called *cap tube,* is the simplest refrigerant flow control device used in refrigeration systems. A cap tube is a small-diameter tube having a calibrated bore through which refrigerant flows into the evaporator. The cap tube is not adjustable and cannot be readily regulated. It is used only on flooded systems and allows the liquid refrigerant to flow into the evaporator at a predetermined rate, which is determined by the size of the refrigeration machine and the load it must carry. The inside diameter (ID) and length of the cap tube are determined by the type of refrigerant, the unit capacity, and operating temperature.

The inlet of the cap tube is high-pressure liquid refrigerant and the outlet is low-pressure liquid refrigerant.

EVAPORATOR

The evaporator is that part of the system where low-pressure liquid refrigerant vaporizes as it picks up heat. Refrigerant leaving the evaporator should be all low-pressure vapor. If too much refrigerant is metered into the evaporator, it is said to be flooded. A flooded evaporator will not cool well because its pressure is too high for it to boil. A flooded evaporator will allow liquid refrigerant to leave the evaporator; this could cause serious damage to the compressor.

If too little refrigerant is metered into the evaporator, the system is said to be starved. The system does not cool properly because the refrigerant will boil off too rapidly before it passes through the evaporator.

Ideally, liquid refrigerant should vaporize off two-thirds to three-quarters of the way through the evaporator. At this point, the refrigerant is said to be saturated, having picked up all the latent heat required to change from a liquid to a vapor without undergoing a temperature change.

From the saturation point the refrigerant vapor will pick up additional heat before leaving the evaporator and a slight amount of heat in the suction line. This added heat, called *superheat*, does not result in a pressure increase.

Suction Line

The suction line is the tube between the evaporator and compressor. An accumulator may be found in the suction line. The suction line may also have a muffler and/or a drier. The refrigerant in this line is low-pressure superheated vapor.

Accumulator

The accumulator, a tanklike vessel, is located near the inlet of the compressor in the suction line. If there is more liquid refrigerant in the evaporator than can be evaporated the accumulator will prevent excess liquid from entering the compressor.

The accumulator allows the low-pressure refrigerant vapor to pass on to the compressor and traps low-pressure liquid refrigerant and oil. A calibrated orifice, known as an "oil bleed," is found in the outlet of the accumulator and allows a small amount of liquid to return to the compressor with the vapor.

Lines

Low- and high-pressure refrigerant fluid and vapor lines may be made of copper, steel, or aluminum. Although tubing sizes vary from system to system, the condition of the refrigerant at various points in all systems is basically the same.

Muffler

Some systems may have a muffler in the suction and/or discharge lines. The condition of the refrigerant in a muffler is the same as in the line to which it is attached.

Summary

The refrigeration cycle exhibits several processes as the refrigerant changes state: liquid to vapor and vapor to liquid.

- When the pressure of the refrigerant drops in the evaporator, the refrigerant boils and, while boiling, picks up heat.
- The compressor increases the temperature and pressure of the refrigerant so that it will condense in the condenser.
- In the condenser, the refrigerant vapor becomes a liquid when giving up the same heat (in British thermal units [Btu]) that was picked up in the evaporator.
- The metering device controls the flow of liquid refrigerant into the evaporator, which separates the high side of the system from the low side.
- The compressor increases refrigerant vapor pressure and thereby separates the low side of the system from the high side.

This is the basic air-conditioning circuit from which all refrigeration circuits are patterned.

Throughout the system, the temperature and pressure of the refrigerant have a direct relationship, one with the other. For example, if the pressure of the refrigerant is known, its temperature can be determined. Conversely, if its temperature is known, the refrigerant pressure can be determined. These determinations are made possible for various refrigerants using a temperature-pressure chart, similar to that found in Table 1 and on the inside back cover.

BE TEST WISE—PASS THE TEST

If you are well prepared to take the test and are familiar with test-taking strategies, you should have no problem passing the first time.

Introduction

TABLE 1: Temperature/Pressure Chart for Refrigerants

TEMP (°F)	VAPOR PRESSURE (psig)										
	R-11	R-12	R-22	R-134a	R-401a	R-401b	R-402a	R-402b	R-404a	R-500	R-502
−30	*27.8	*5.5	4.9	*9.7		*7.3	11.5	9.0	10.5	*1.4	9.4
−25	*27.5	*2.3	7.2	*6.8		*4.2	14.8	12.0	13.6	1.1	12.3
−20	*27.0	0.6	10.2	*3.6		*0.6	18.3	15.4	16.6	3.1	15.5
−15	*26.6	2.4	13.2	*0.2		1.6	22.2	18.6	20.7	5.4	19.0
−10	*26.0	4.5	16.5	2.0	1.9	3.8	26.4	22.6	24.2	7.8	22.8
−5	*25.4	6.7	20.0	4.2	4.0	6.1	31.0	27.0	28.4	10.4	26.8
0	*24.7	9.2	24.0	6.5	6.4	8.7	36.0	31.0	33.0	13.3	31.2
5	*24.0	11.8	28.2	9.1	9.0	11.5	41.3	36.0	38.5	16.4	36.0
10	*23.1	14.6	32.8	11.9	11.8	14.5	47.1	42.0	43.3	19.8	41.1
15	*22.1	17.7	35.7	15.1	14.8	17.8	53.3	47.0	50.0	23.4	46.6
20	*21.1	21.0	43.0	18.4	18.1	21.4	60.0	54.0	52.2	27.3	52.5
25	*19.9	24.6	48.9	22.0	21.7	25.3	67.2	60.0	63.4	31.6	58.7
30	*18.6	28.5	54.9	26.1	25.6	29.5	74.8	67.0	68.9	36.1	65.4
35	*17.2	32.6	61.9	30.4	29.8	34.1	83.0	75.0	78.7	41.0	72.6
40	*15.6	37.0	68.5	35.1	34.4	39.0	91.8	83.4	84.6	46.2	80.2
45	*13.9	41.7	76.0	40.1	39.2	44.3	101.1	91.5	98.2	51.9	88.3
50	*12.0	46.7	84.0	45.5	44.5	49.9	111.0	100.0	102.4	57.8	96.9
55	*10.0	52.1	92.6	51.3	50.0	56.0	121.5	110.0	114.6	64.2	106.0
60	*7.7	57.7	101.6	57.3	56.1	62.5	132.7	120.0	125.1	71.0	115.6
65	*5.3	63.8	111.2	64.1	62.5	69.4	144.6	132.5	136.3	78.2	125.8
70	*2.7	70.2	121.4	71.2	69.3	76.8	157.1	143.4	148.2	85.8	136.6
75	0.1	77.0	132.2	78.7	76.6	84.7	170.4	155.0	160.1	93.9	148.0
80	1.6	84.2	143.6	86.8	84.4	93.1	184.4	170.0	174.0	102.5	160.0
85	3.2	91.8	155.7	95.3	92.6	102.0	199.2	182.5	188.1	111.6	172.5
90	4.4	99.8	158.4	104.4	101.4	111.4	214.7	197.5	202.9	121.2	185.8
95	6.9	108.3	181.8	114.0	110.7	121.4	231.2	213.3	218.6	131.3	199.7
100	8.9	117.2	195.9	124.2	120.6	132.0	248.4	230.0	235.1	141.9	214.4
105	11.1	126.6	210.8	135.0	131.0	143.3	266.6	246.7	252.5	153.2	229.7
110	13.4	136.4	226.4	146.4	142.0	155.1	285.7	263.3	270.8	164.9	245.8
115	15.9	146.8	237.7	157.5	153.6	167.7	305.7	283.3	290.0	177.4	262.6
120	18.5	157.7	259.9	171.2	165.9	180.9	326.7	303.3	310.3	190.3	280.3
125	21.3	169.1	277.9	184.6	178.8	195.8	348.7	323.3	331.3	203.9	298.7
130	24.3	181.0	296.8	198.7	192.5	209.5	371.7	345.0	353.9	218.2	318.0

NOTES: − denotes below zero temperature (°F)
       * denotes vacuum pressure (inches of mercury)
      (1) All pressure rounded to nearest 0.1 psig or inches of mercury.
      (2) For absolute (psia) pressure, add 14.7 to pounds per square inch–gauge.
      (3) For metric gauge pressure (kPa), multiply pounds per square inch–gauge by 6.895.
      (4) For metric absolute pressure (kPa absolute), multiply pounds per square inch–absolute by 6.895.
      (5) For °C temperature, subtract 32 from °F and then multiply by 0.555.

BEFORE TEST DAY

Here are a few tips that you may consider before test day.

- Study the technical material relating to your test.
- Read over the material and make an estimate of how much time you think will be required for you to absorb the information.
- Prepare a schedule to fit your study time into the time left before the test date. Some suggest that you double the time estimated so you are fully prepared. For example, if you decide that 20 hours of study are needed, and there are 10 days before test day, you should plan for study four hours each day.
- Adjust your daily activities so that the time of day you study is about the same every day.
- Recognize that the extent of your determination to master the material and pass the test is a very important factor in your success.
- Keep a record of the time you spend studying. If you keep a record, you will be more inclined to study daily.

TEST-TAKING FEAR

A healthy anxiety or fear about taking a test is generally a good thing. At test time, some anxiety will pique your senses and provide a healthy flow of adrenaline through your system.

Do not, however, become overly concerned about taking the test. Remember, you have been tested many times before. Life itself is a continual test. The pending test is just that—another test. If you allot sufficient time, follow your schedule, and study effectively, you should do well.

Tests can be frightening. Not many people look forward to taking a test. Although you may be afraid of tests, it doesn't mean that you have to fear them. A few test-taking techniques should take the fear out of them.

Before learning about test-taking techniques, however, we'll cover one of the key problems many face: test anxiety. Test anxiety is a reaction to a test, characterized by sweaty palms accompanied by a blank mind.

If someone were to tell you that they don't test well, they are really saying that they do not study or prepare well. For some, it could mean they are easily distracted or that they are not mentally or emotionally prepared.

Introduction

Most recognize the competitive nature of tests. Being in the right frame of mind when taking a test is important. Some will rise to the occasion when facing such a challenge but others will be thrown off balance by the pressure. Actually, either reaction could have little to do with one's knowledge, intelligence, or preparation.

One of the best ways to avoid test-taking pressures is to place yourself in the test-taking environment as often as possible by practice testing.

RELAX

If your mind is crammed with facts and figures, names, and dates, you may find it difficult to concentrate on specific details that you need to recall. This may be true even if you know all the material being covered in the test. The adrenaline rushing through your body may make instant retrieval near impossible.

The simplest relaxation technique is deep breathing. Just lean back in your chair, relax your muscles and take three very deep breaths, counting to 10 while holding each breath and exhaling slowly. For many, that's the only relaxation technique needed.

MEDITATE

Meditation techniques are based on a principle similar to that of relaxation. Focus your mind on one thing to the exclusion of everything else. While you're concentrating on the subject of your meditation, your mind can't think about anything else.

The next time you can't concentrate, sit back, take three deep breaths, and concentrate for a minute or two on a word, any word. When you're done, you should be in a more relaxed state and ready to shift your concentration to the test. The more you believe in this technique, the better it will work for you.

DISTRACTIONS

You must be able to concentrate when studying. Outside influences should be kept to a minimum and, better yet, completely eliminated. Make sure that you are not distracted by light or noise. Some technicians listen to soft music while studying; others are better off with complete quiet. Very few people, if any, can really study while the television is on.

In order to study effectively, you must be well rested. You should get about eight hours of sleep each day when preparing for the test. If possible, you should not study immediately prior to retiring for the night. People are normally least alert just before bedtime.

If you become stuck on the meaning of a word during your study, look it up in the glossary of your reference text. If the word is not there, use a dictionary. Do not make the mistake of guessing at a word, because it might change the whole meaning of a question or an answer.

CRAMMING

It is human nature to wait until the last minute and then try to cram a week's worth of studying into a single night or weekend. Cramming works for only a few technicians. These few are able to store more information in short-term memory than most of us can. Also, they can and actually remember it for about 24 hours, or until after the test.

Most of us, however, do not do well on a test after a night of no sleep and too much study. In fact, we're lucky if we can remember what the test instructions were. That's the best reason not to cram. For most of us, it just does not work!

In spite of your best resolve, you may find that it is necessary to do some studying the night before a test. If that is the case, there is only one rule to follow to help make your night of cramming marginally successful: Know when to give up.

TEST YOURSELF

For practice, answer the sample questions in this book. You may even construct your own tests. The more you practice, the better prepared and more confident you will be when taking the actual test. If a co-worker is also taking the test, ask each other the questions.

Practice tests offer some real advantages. Familiarization with whatever type of test you're taking is vitally important. It enables you to strategically study and attack the test. Also, familiarization breeds comfort, and being comfortable is a key ingredient to doing well.

GROUND RULES

Are you penalized if you guess? The proctor for some tests, for example, may tell you that you earn two points for every correct answer but lose one point for every incorrect one. This is not the case, however, with EPA's refrigerant certification tests.

Some tests may have two or more sections. Some sections may count only 10 or 20 percent of your final score, while another section may account 50 percent or more of your final score. This is not the case with the refrigerant certification test. These test questions are not weighted—all questions have the same scoring value.

DISCRIMINATE AND ELIMINATE

There is no penalty for guessing. Because there is no penalty for a wrong answer, you should never leave an answer blank. You should, however, do all that you can to increase your odds of getting the correct answer. Multiple-choice questions have four choices. If you guess, you have a 25 percent chance of being correct. Of course, there is also a 75 percent chance of being wrong. If you don't guess, however, the odds are greater: 100 percent that you are wrong.

If you are able to eliminate one of the choices but are not sure of the correct answer, you have increased your odds of being right to 33 percent. If you are able to eliminate two of the choices but are still not sure of the correct answer, you have increased your odds of being right to 50 percent. Here are a few real insider tips to help make your guess more educated.

- Do not read too much into the question. Do not try to second-guess the test writer.
- Underline the key words if you are permitted to write on the test paper.
- If two choices are very similar, the answer is probably neither one of them.
- If two choices are very close in meaning, choose one of them.
- If two choices sound alike, choose one of them.
- If two choices are opposite, one of them is probably correct.

- Do not go against your first impulse unless you are sure you were wrong.
- Check for negatives and other words that are there to throw you off, such as "Which of the following is NOT . . . ." or "All of the following EXCEPT . . . ."
- An answer is usually wrong if it contains the word "all," "always," "never," or "none."
- An answer is probably correct if it contains the word "sometimes," "probably," or "some."
- When the correct choice is unknown, try to eliminate the incorrect choices.
- Do not eliminate an answer choice unless you know what every word means.
- Read every answer choice (unless you are guessing in the final minutes of the test).
- Make sure you mark each answer in the correct place.
- The longest and/or most complicated answer to a question is often the correct answer. The test writer has added qualifying phrases to make that choice correct.
- Be suspicious of choices that seem too obvious.
- Don't give up on a question that, after one reading, seems hopelessly confusing or difficult.
- If there is a mathematical question, choose the answer in the middle.
- If two quantities are very close, choose one of them.
- If two numbers differ only by a decimal point, and the others are not close, choose one of them.
- If two choices to a mathematical problem look alike, either in formula or shape, choose one of them.
- If you are required to read a passage before answering a question about it, read the question first. That will tell you what you're looking for in the passage.
- If choice (4) or (D) is "All of the above," it is often the correct answer. You do not have to be sure that "all of the above" is correct to choose it. All you have to be is pretty sure that two choices are correct, or they are not wrong.
- If choice (4) or (D) is "None of the above," it is often the wrong answer. You have to be sure that "none of the above" is correct to choose

it. You have to be pretty sure that all other choices are wrong, or sure they are not correct.

CHANGE AN ANSWER

Should you go back, recheck your work, and change an answer? No! Statistics show that your first guess was pretty good, presuming you had some valid basis for guessing. Change it only if you are sure your first guess was wrong.

TEST-TAKING TECHNIQUES

There are three very specific techniques that you may use for answering multiple-choice-type questions. One of the techniques might work for you.

Technique 1

Start with the first question and continue with the second, then the third, and keep going, question by question, until you reach the last question. Never leave a question until you have answered it or, at least, made an educated guess.

This technique is perhaps the quickest, because no time is wasted reading through the whole test trying to pick the easiest or hardest questions. Do not, however, allow yourself to get so stumped by a single question that you spend too much time on it. This is more than likely the technique most test takers will use.

Remember, however, to skip questions that are confusing for you and then go back to them, and, if you must, guess. Be sure to allow enough time to go back to those you haven't answered and to check all your answers.

Technique 2

Answer every easy question. Answer those that you are sure of or those that require the simplest calculations first, then go back and answer the harder questions.

This technique ensures that you will maximize your right answers. You are answering those that you are certain of first. This approach may also give you the most time to work on those that you find difficult.

Many experts recommend this technique because they feel that answering so many questions one after another gives you immediate confidence to answer the questions that you are not sure about.

Technique 3

Answer all the hardest questions first, then go back and answer the easy questions. If the pressure of time starts getting to you near the end of the test, you will be in a position to answer a lot of the easiest questions in the limited time remaining, rather than facing those that you may really have to think about. By the end of a test your mind will not be working as well as it was at the beginning.

This is not the technique for everyone, but it may be right for you. Try to ensure adequate time to answer every question. It is better to get one answer wrong and complete three other answers than get one right and leave three blank.

None of these options are right or wrong. Each may work for different individuals, assuming that these approaches are all in the context of the test format.

And don't fall into the "answer daze"—the blank stare some test takers get when they just can't think of an answer. It is better to move on and get that one question wrong than waste valuable time doing nothing.

READ AND UNDERSTAND

A key point of preparation is to read and understand the directions. Otherwise, you could do everything correctly but not follow your proctor's directions, in which case everything is wrong.

Look through the entire test and break it down into time segments. Count the total number of questions, divide by the time allotted, and set goals on what time you should reach question 10, question 20, and so on.

If there are facts or formulas you are afraid you'll forget, it is a good idea to write them down somewhere in your test booklet before you start. It won't take much time and it could save some serious memory jogs later.

TEST-TAKING STRATEGIES

There are a number of strategies that can be used for answering multiple-choice questions. They are given in the order suggested for use during a test. For example, if strategies 1 through 3 provide the answers, it is not necessary to use strategies 4 through 9.

Strategy 1—Work Quickly

Studies have shown that those who rapid-fire their way through multiple-choice tests, even if they pick some random answers, get better scores than those who may have a better knowledge of the subject matter but are slow at taking tests.

Read each question through once. Do not hesitate to put down the correct answer the instant you come to it. If you have a second thought about the correct answer, do not stop to think about it. Place a little mark alongside the question number. If you have time at the end of the test, you can reconsider your first answer.

If an answer seems obvious, have confidence in yourself. Choose the simple answer. Do not waste time looking for hidden qualifications and tricks.

Strategy 2—Give the Answer Required

Learn to read and understand test instructions as well as the questions. Failure to do so can cost points on your score.

Make sure that you understand precisely what the directions tell you to do. Some test questions ask for the *most* correct answer. In that case, you may have to decide on more than just the answer. It is not what seems most correct to you that counts, but what the industry believes is most correct.

For example, you may think the ozone-depletion story is a fairytale. The industry, on the other hand, feels that ozone depletion is a serious problem. When taking the test it is not the time to take a stand on your own opinions.

Study the given part of the question, called a *stem*. If the stem includes several statements, isolate each of them and make sure that you understand each individual part. When you have selected an answer, check it against each segment. Your answer should satisfy every part of the question. Breaking down complicated questions into smaller ones rewards test takers with extra points on almost every test.

Strategy 3—Guess Before You Choose

Read the question and try to decide on your answer before looking at the choices given. If the answer you thought of is among the choices, you have saved a lot of time. If it is not one of the choices, clear your mind and study each option. You have at least activated the part of your memory that applies to the topic. That can be great in helping to recognize the correct answer.

Strategy 4—Select the Closest Answer

By their nature, multiple-choice questions are simplistic, so look for a quick and easy response. For that reason, choose the closest answer, even if you do not think that it is 100 percent correct.

If a question is well constructed, all the answer choices will seem relatively plausible. If that is the case, there will probably be at least one clue word in the stem or question that will make one answer better than the other three choices. Reread the question and stem if necessary and look for that clue word.

Strategy 5—Eliminate Improbable Answers

The more answer choices you can eliminate, the greater your odds of making the correct choice. Some choices are obviously wrong. If that is the case, quickly move to the next choice. Many choices are partly wrong. If a choice is wrong in any way, it is not the correct choice.

If you think that a question is badly written or that the answer choices are inappropriate, report it to the proctor after you have completed the test.

Many of the choices are correct statements by themselves, but may have nothing to do with the stem part of the question.

Strategy 6—Look for Clues

You cannot always decide on the correct answer by looking for clue words. If the first five strategies don't work, give this one a try. Looking for clues may lead you to a choice that you suddenly remember to be correct.

Look for absolutes and qualifiers. Choices that include *"always," "never," "all,"* and *"none"* are often incorrect. Few things in life are *always* true or *always* false. Test writers like to avoid arguments over the answers by using qualifiers such as *"seldom," "generally,"* and *"tend to be."*

Look for grammatical clues. The tendency among test writers is to have the correct answer agree grammatically with the stem or question. Seldom do test writers take such care with the incorrect responses.

Look for familiar phrases. Test stems, questions, and correct responses are often very similar to those in study guides, textbooks, or notes. If you recognize particular words or phrases, or if the stem and one choice flow smoothly together in your mind, follow your intuition. It is probably correct.

Look for a degree of correctness. If the answer required is a number, one choice is likely to be too large and one is likely to be too small. The correct choice, then, is probably one of those in the middle. If a date is involved, one choice is likely to be too early and one choice is likely to be too late. Again, the correct choice is probably one of the two in the middle.

If one possibly correct choice is very specific, and another possibly correct choice is very general, the general choice may be the correct one. This is usually true if the general choice comprises all, or most, of the information in the specific choice.

While none of these suggestions are certain, they are a lot more dependable than just taking a guess, the subject of our next strategy.

Strategy 7—Guess

On a four-part multiple-choice test, random guessing all the questions should result in an average score of 25 percent. On four-part multiple-choice tests, you should be able to eliminate two choices by using what you know by incorporating the strategies presented in this guide. That can result in an average score of 75 percent achieved by guessing alone. If you use a bit of intelligence, you can push the guess rate beyond 80 percent. Not bad when you consider that the passing score is 70 percent.

If all strategies fail and your mind is a total blank on the tests, you still may be able to slant the odds in your favor by looking for patterns in the answers. If, for example, choice (C) has not been chosen on your answer sheet for a long time, fill in (C). Any answer is better than no answer.

Strategy 8—Change the Answer

A popular old wive's tale is "Never change an answer." Do not subscribe to this saying. Studies now indicate that when you have an intuition that you ought to change an answer, your intuition usually proves to be right.

Do not review your answers until after you have finished the test. After that, use all the remaining time that you have for review.

First, reread the directions. Make sure that you have followed them to the letter. Next, make sure you have put all your answers in the correct places. Many test takers lose points for simply not checking this. Check over the answers, if any, that you marked for further consideration.

Finally, if you still have time, go over your other answers. If you believe you should change an answer, change it. Often just answering all the questions may reveal clues, consciously or subconsciously. By the end of the test you should have a better understanding of the objectives of the test.

Do not, however, change answers back and forth. Repeated changes of the same answer rarely pays off. Before handing in your test papers, be sure to erase all marks that do not belong on them.

Strategy 9—Do Not Give Up

Many test takers lose points because they give up before working their way through the eight strategies. Test-wise test takers keep moving along rapid-fire, tackling each question in turn while looking for the key words that count and guessing at probable answers as they read the stem.

Some test takers search the choices quickly for the preferred answer, systematically eliminating unlikely answers so that they can make an educated guess from the others, and attempt to answer each question before moving on.

The odds are in your favor if you keep going until the end.

ON TEST DAY

On test day, try to observe the following rules:
- Be sure to get plenty of rest the night before the test.
- Arrive at the test site on time. Early is better than late. Take into account possible traffic or other delays.
- Have with you whatever materials you think that you need to take the test. This may include pencils, a watch, a pocket calculator, a temperature/pressure chart, and a candy bar (if permitted).
- Go to the test room as early as possible so that you can select a good seat.

Introduction 25

- Choose a seat away from windows. You do not want to be distracted or waste your time looking out a window.
- Listen to the proctor's instructions and follow them exactly.
- When told to begin, quickly read through the test and answer the questions that are obvious to you.
- Answer all the questions in the test. Select the answer that is your best guess if you are not sure of the correct answer. Remember, an unanswered question is wrong. By guessing, you have at least a 25 percent chance of being correct.
- If you finish the test before the time limit, review the questions and your answers. Errors are sometimes made in marking an answer. Now is the time to make corrections.
- Continue to review your work until time is called. If you have followed a good schedule and study procedure, are well rested, and concentrate, you will pass the test!

GOOD LUCK

The practice questions should give you a good insight of what to expect on the test. For ease of study they are broken into four categories: Basic, Type I, Type II, and Type III. Universal Certification is granted when a technician has passed all the tests.

A glossary is provided in the Appendix for some terms with which you may not be familiar.

HEALTH AND SAFETY

You should understand the importance of the ozone layer and how it is created and destroyed. You should also understand what the industry and the government is doing about ozone depletion.

WHAT IS OZONE?

A leading dictionary describes ozone as having a penetrating refreshing odor while providing an exhilarating influence. Ozone is a molecular form of oxygen having a different chemical property, an *allotrope* of oxygen. In heavy concentrations, ozone is a poisonous gas. The ozone layer, on the other hand, protects animal and plant life from damaging ultraviolet (UV) radiation.

To most people, ozone has a pungent odor described by many as irritating. In high concentrations, ozone has a pale blue color in contrast with oxygen, which is colorless, tasteless, and has no odor.

ATMOSPHERIC OZONE

Most other gases are concentrated in the troposphere. About 90 percent of the ozone (O_3), however, is found in the stratosphere, between about 9 and 22 miles (15 and 35 km) above the earth. Even at its greatest concentration ozone does not exceed 10 ppmv: this means there is only one ozone molecule in every 100,000 molecules. If all the ozone in the atmosphere, for example, were concentrated at sea level, it would form a layer less than 0.125 inch (3 mm) thick. There are about 3000 million tons of ozone in the atmosphere, equivalent to about 1600 pounds (726 kg) per person on Earth. Compared with the mass of the atmosphere, however, the amount of ozone is negligible.

Ozone is formed by the action of electrical discharges. This may explain why it is sometimes detected by odor near electrical equipment or after a thunderstorm. More frequently, however, it is formed by the action of UV radiation on oxygen (O) in the stratosphere. The oxygen atoms in the oxygen molecules split apart and the separated atoms recombine with other oxygen molecules to form the triatomic ozone (O_3).

Sunlight is essential for the formation of stratospheric ozone. It is, therefore, mainly formed over the equatorial region, where solar radiation is the greatest. From there it is distributed throughout the stratosphere by the slight global wind circulation. Stratospheric ozone levels vary throughout the world, being lowest at the equator and highest toward the poles.

ABSORPTION OF ULTRAVIOLET RADIATION

Radiation from the sun is of various wavelengths that range from UV through visible light to infrared (IR) light. Ultraviolet radiation can be very damaging to living organisms, causing sunburn, skin cancer, damage to eyes including cataracts, and premature aging and wrinkling of the skin. It can also break the food chain by destroying minute organisms such as plankton in the ocean, thereby depriving certain species of their natural food. Plant life and crops can also be devastated by excessive UV radiation.

The damaging forms of UV radiation are absorbed by ozone in the atmosphere and do not reach the earth. A small amount of ozone in the atmosphere is sufficient to absorb this radiation and acts as a giant sunscreen or umbrella enveloping the earth. Depletion of the ozone allows more UV radiation to strike the earth and its living organisms.

Another consequence of the absorption of solar energy by ozone is that the upper stratosphere is somewhat warmer than at lower altitudes; this helps to regulate the earth's temperature. Stratospheric ozone absorbs about 3 percent of incoming solar radiation, thus serving as a heat sink. Loss of ozone will decrease the temperature of the stratosphere, which will, in turn, affect the troposphere and, consequently, the weather and climate at the earth's surface.

DESTRUCTION OF THE OZONE

Ozone is both created and destroyed by the action of UV radiation on oxygen molecules. Chlorine (Cl) is the major gas causing the destruction of ozone and starts chain reactions in which a single molecule of chlorine can destroy 100,000 ozone molecules over time. Such reactions can continue for many years, even a century or more, until the chlorine drifts down into the troposphere or is bound into another compound.

The main sources of chlorine are chlorofluorocarbons (CFCs), also referred to as Freon and halons. Chlorofluorocarbons are artificially made chemicals first developed in 1928 and comprised of:

- Chlorine (Cl)
- Fluorine (F)
- Carbon (C)
- Often, hydrogen (H)

They are very stable chemicals and are nonflammable, nonirritating, nonexplosive, noncorrosive, odorless, and relatively low in toxicity. They vaporize at low temperatures, which makes them very suitable for use as a coolant in refrigerators and air conditioners, as a solvent in cleaning electronics, in blowing bubbles in certain types of foam-blown plastics such as sponge plastic and food packaging, and in cleaning solvents for dry cleaning. Halons, mainly used in fire extinguishers, will not be discussed in this text.

The consumption of CFCs in the United States on a per capita basis is among the highest in the world, a reflection of our affluence and the popularity and use of air conditioners.

In 1974 two chemists at the University of California, Mario Molina and Sherwood Rowland, asked a simple question: "What has happened to the millions of tons of CFCs released over the previous four decades?" The only "sink" they could suggest was the stratosphere. They hypothesized that the chemical stability of CFCs would enable them to reach the stratosphere, be broken apart by the intense UV radiation and release chlorine by a process known as photolysis. The Cl would then react with the ozone causing its depletion.

It is not the CFCs that cause the destruction but rather the chlorine released by the CFCs. The research of the British scientists at Halley Bay during the 1980s, together with international research programs in late 1987 in which samples of stratospheric air were obtained by high-altitude flights over Antarctica, proved the link between CFCs and ozone destruction.

A chlorine atom reacts with an ozone molecule, splitting it apart and attaching itself to one of the oxygen atoms to form chlorine monoxide. A free oxygen atom splits the chlorine monoxide molecule to re-form a molecule of oxygen, and the chlorine atom is free to attack another ozone molecule.

The CFCs and halons take six to eight years to rise from Earth up through the atmosphere. Chlorine as used in swimming pools and chlorine bleach, on the other hand, is unstable and breaks down rapidly without rising into the atmosphere. The concern is that the current ozone hole and depletion have resulted from CFCs released in the early 1980s; their continuing high levels of use through the 1980s and early 1990s have yet to make their presence felt in the stratosphere.

EFFECTS OF OZONE LOSS ON HUMAN HEALTH

To protect life on Earth from damaging UV radiation, ozone acts as a giant sunscreen, absorbing the UV rays. This prevents a certain percentage of UV radiation from reaching Earth. Loss of ozone will thus allow more UV radiation to penetrate Earth and this will adversely affect human health and the environment. The three areas of our bodies that are adversely affected are the skin, eyes, and the immune system.

Exposure of skin to UV radiation can initially result in sunburn and suntan, and if the exposure continues over a long period, as with those who work outdoors, the skin protects itself from UV radiation by gradually thickening and darkening as a pigment called melanin is released in the skin. Continual exposure of the skin to UV radiation results in its aging and wrinkling, and this increases the risk of skin cancer.

Excessive UV exposure to the eyes will increase the risk of cataracts, which cause cloudiness in the lens of the eye, limiting vision. Other eye problems such as retina damage, tumors on the cornea, and "snow blindness" may also be caused by exposure to increased levels of UV radiation.

The body's immune system protects it from foreign chemicals and infections. If damaged, the immune system cannot protect the body, and infections spread more rapidly. UV radiation reduces the ability of the immune system to reject cancers, although not much is known about why this happens. Overall, increased UV radiation resulting from ozone depletion has the potential of significantly increasing human skin cancers and cataracts and damaging the human immune system. It also adversely affects marine and terrestrial plants and animals.

The extent of the damage will depend on the degree by which the ozone layer is depleted. To date it has been reduced by about 2.5 percent and it remains to be seen if the action taken to control the release of ozone-depleting substances will be sufficient.

THE GREENHOUSE EFFECT

The greenhouse effect is a natural process of warming, just as the ozone layer is a natural function of the earth's atmosphere that protects life. Both have been affected by the release of pollutants by human activities, pollutants that have accelerated the greenhouse effect, thus resulting in increased warming, and depletion of the ozone layer, exposing life to damaging UV radiation.

THE CLEAN AIR ACT

The most significant legislation to affect the industry in the United States is the Clean Air Act (CAA). The CAA was signed into law by President George

Bush on November 15, 1990. Most of the rules and regulations of the CAA were the result of the recommendations made at the Montreal Protocol.

The Montreal Protocol and, later, the Copenhagen Amendments, deals with the environmental problems and issues created by certain refrigerants depleting the ozone on an international level. The CAA deals with this problem on a national level. The Montreal Protocol is structured so that periodic meetings must take place in order to reassess the ozone problem. As new facts about the impact of refrigerants are reported, the Protocol will be modified accordingly. The majority of Protocol modifications will also result in the CAAs being modified accordingly.

Language exists in the CAA stating that the EPA can accelerate phaseout schedules if it is deemed necessary and practical. The CAA also mandates that phaseout may be accelerated if required by the Montreal Protocol.

The CAA is somewhat more specific than the Protocol in addressing the ozone-depletion problem. The Act gives the EPA the authority to establish environmentally safe procedures with respect to the use and reuse of refrigerants. In addition, the EPA will establish standards for certification and service of refrigeration equipment and those who service the equipment. These standards will be derived from the information furnished mainly by private sector organizations.

STRATOSPHERIC OZONE PROTECTION—TITLE VI

The CAA has a section called Title VI—Stratospheric Ozone Protection, which establishes regulations for the production, use, and phaseout of CFCs, halons, and HCFCs. Other chemicals such as carbon tetrachloride (CCl_4), also covered by Title VI, are not covered in this text. Title VI breaks the substances to be regulated into two classes—Class I and Class II.

The chemicals that we are primarily concerned with are Class I refrigerants. These refrigerants were scheduled to be phased out in short order.

SIGNIFICANT NEW ALTERNATIVES POLICY

Final Rule Summary

Section 612 of the CAA requires that the EPA establish a program to identify alternatives to Class I and Class II ozone-depleting substances and to publish lists of acceptable and unacceptable substitutes. On promulgation of the final rule, it will be illegal to replace a Class I or Class II substance with any substitute that the Administrator determines may present adverse effects to human health or to the environment. The final rule also makes it illegal to use substitutes other than those that have been identified to reduce overall risk and are currently or potentially available.

The Final Rule

On March 18, 1994, the EPA published the Final Rulemaking (FRM) (59 FR 13044), which described the process for administering the SNAP program and issued EPA's first acceptability lists for substitutes in the major industrial use sectors. These sectors include refrigeration and air conditioning; solvents; foam blowing; fire suppression and explosion protection; sterilants; aerosols; adhesives, coatings, and inks; and tobacco expansion. To assess the acceptability of a substitute, the EPA completes a screening analysis in which overall risks to human health and the environment in use-specific applications are examined.

Updates

The EPA intends to publish updates to the Significant New Alternatives Policy (SNAP) rule. These updates may be in two parts, a "Notice" or a "Rulemaking." A "Notice" contains no regulatory controls and thus does not need to go through the public comment process. It enters into force after publication in the *Federal Register*.

A Rulemaking requires a public notice-and-comment process, beginning with a "Notice of Proposed Rulemaking" (NPRM). An NPRM contains proposed lists of agents deemed "Acceptable Subject to Use Conditions," "Acceptable Subject to Narrowed Use Limits," and "Unacceptable."

There is a 30-day public comment period beginning on the date of publication in the *Federal Register,* after which a "Final Rulemaking" is published. Several notices and rules have been published since the inception of the SNAP program.

Who Must Apply?

Under SNAP, the EPA defines a "substitute" as any chemical, product, substitute, or alternative manufacturing process, whether existing or new, that could replace a Class I or Class II substance. Anyone who produces a substitute must provide the Agency with health and safety studies, as well as notify the Agency at least 90 days before introducing it into interstate commerce for use as an alternative. This requirement applies to chemical manufacturers, but may include importers, formulators, or end users when they are responsible for introducing a substitute into commerce.

To develop the lists of acceptable and unacceptable substitutes, the EPA must assess and compare "overall risks to human health and the environment" posed by use of substitutes in the context of particular applications. The EPA requires submission of information covering a wide range of health and environmental factors. These include intrinsic properties such as physical and chemical information, ozone-depleting potential, global warming potential, toxicity, and flammability, and use-specific data such as substitute applications, process description, environmental release data, environmental fate and transport, and cost information.

The EPA has identified three mechanisms for revising or expanding the list of SNAP determinations published in the final regulation:

1. As new substitutes are developed for commercial sale, producers must notify the EPA at least 90 days before introduction of a chemical into interstate commerce for significant new use as an alternative to Class I or Class II substances;
2. Any person may petition the EPA to add or delete substances from the SNAP lists of acceptable and unacceptable alternatives; and
3. The EPA may revise the SNAP lists outside the context of petitions or notifications, based on new data or on characteristics of substitutes previously reviewed.

Section 2

CERTIFICATION

CORE TEST

GENERAL KNOWLEDGE

Anyone taking the certification exam must pass the CORE test section and at least one other test to be certified. The following information should be helpful when answering the CORE questions.

OVERVIEW OF ISSUES ON THE CORE TEST

The following is an overview of the material that may be on the CORE test.

Ozone Depletion

- ✔ Be aware of the destruction of the ozone caused by chlorine and the presence of chlorine in CFC and hydroclorofluorocarbon (HCFC) refrigerants.
- ✔ Be able to identify CFC, HCFC, and hydrofluorocarbon HFC refrigerants. There is no requirement that you know the chemical formulas, but you should know, for example, that R-12 is a CFC, R-22 is an HCFC, that R-134a is an HFC, and so on.
- ✔ Have knowledge that CFCs have a higher ozone-depletion potential (ODP) than HCFCs, which have a higher ODP than do HFCs. Be aware of the health and environmental effects of ozone depletion and of the

Certification

evidence of ozone depletion and the role that CFCs and HCFCs play in ozone depletion.

Clean Air Act

- ✔ Be able to recall the CFC phaseout dates.
- ✔ Be aware of the venting prohibition during equipment servicing as well as during equipment disposal and the rules regarding substitute refrigerants beginning in November 1995.
- ✔ Be able to recall the maximum penalty for venting refrigerants under the CAA.

Section 608 Regulations

- ✔ Be able to recall the definition and/or identification of high- and low-pressure refrigerants and the definition of system-dependent versus self-contained recovery/recycling equipment.
- ✔ Recall the identification of equipment covered by the rule pertaining to all air-conditioning and refrigeration equipment containing CFCs or HCFCs, except motor vehicle air conditioners (MVACs).
- ✔ Understand the need for third-party certification of recycling and recovery equipment manufactured after November 15, 1993.
- ✔ Be aware of the Air-Conditioning Refrigeration Institute (ARI) 700 Standard for reclaimed refrigerants.

Substitute Refrigerants and Oils

- ✔ Be aware of the absence of so-called "drop-in" replacement refrigerants and the incompatibility of substitute refrigerants with lubricants used with CFC and HCFC refrigerants as well as the incompatibility of CFC and HCFC refrigerants with new lubricants. This includes the identification of lubricants for any given refrigerant, such as esters with R-134a.
- ✔ Be aware of the fractionation problem, the tendency of the different components of a blend refrigerant to leak at different rates.

Refrigeration

- ✔ Know the refrigerant states (vapor vs. liquid) in a system and the pressures at the different points of a refrigeration cycle. Know how, when, and where cooling occurs within the system.

✔ Be familiar with refrigeration test gauges, their color codes, pressure ranges, different types, and proper use.

Three Rs

✔ Be able to define recover, recycle, and reclaim.

Recover: To *recover* refrigerant is to remove it, in any condition, from an appliance and store it in an external container without necessarily testing it or processing it in any way.

Recycle: To *recycle* refrigerant is to extract it from an appliance and clean it for reuse without meeting all the requirements for reclamation. In general, this means that the refrigerant has been cleaned using oil separation and single or multiple passes through devices to remove moisture, acids, and particulate matter.

Reclaim: To *reclaim* refrigerant is to process it to at least the purity specified by ARI Standard 700-1993, "Specifications for Fluorocarbon Refrigerants," and to verify this purity using the methods specified.

Recovery Techniques

✔ Understand the importance of avoiding contamination by the mixing of refrigerants.
✔ Be aware of the factors affecting the speed of recovery, such as ambient temperature, size of recycling or recovery equipment, and hose length and diameter.

Dehydration Evacuation

✔ Understand the need to evacuate a system to eliminate air and moisture at the end of a service procedure.

Safety

✔ Understand the risks of exposure to refrigerant such as oxygen deprivation, cardiac effects, frostbite, and long-term hazards.
✔ Know to use personal protective equipment such as gloves, goggles, and self-contained breathing apparatus (SCBA), in extreme cases.

✔ Be able to distinguish between types of cylinders, such as disposable cylinders by Department of Transportation (DOT) designation, color, and markings and reusable recovery cylinders, gray with yellow top.
✔ Know the hazards that are involved in filling a refrigerant recovery cylinder more than 80 percent full.
✔ Understand the reasons for using nitrogen rather than oxygen or compressed air for leak detection and why a pressure regulator and relief valve must be used with nitrogen.

Shipping
✔ Recognize and identify the labels and classification tags required for shipping refrigerant cylinders required by DOT.

OZONE DEPLETION

For more than 50 years, CFCs and HCFCs have been responsible for dramatically changing our lifestyles. The use and release of these compounds into the atmosphere have had far reaching effects on the earth's environment. The greatest effect is the depletion of ozone in the stratosphere.

The stratosphere, Earth's security blanket, is located 10 to 30 miles (16–48 km) above Earth's surface. It is comprised of, among other things, ozone, referred to as the ozone layer. An ozone molecule consists of three Oxygen atoms (O_3). The ozone layer actually serves two very important functions. (1) it protects us from harmful ultraviolet (UV) radiation and (2) it helps to maintain a stable Earth temperature. Stratospheric ozone depletion is a problem that affects the entire planet. Depletion of ozone in the stratosphere causes crop loss, increased eye diseases, skin cancer, loss of marine life, loss of plant life, and increased ground-level ozone.

Releasing CFCs and hydrochlorofluorocarbons (HCFCs) into the atmosphere depletes the ozone layer. The chlorine in these compounds is the problem. When a chlorine atom meets with an ozone molecule, it is joined by an oxygen atom from the ozone. This action forms a compound called chlorine monoxide and leaves a two-sided oxygen (O_2) molecule behind. The chlorine monoxide then collides with another ozone molecule and releases its oxygen atom, forming two O_2 molecules and leaving the chlorine atom free to attack another ozone molecule. The life span of chlorine in the stratosphere is such that one chlorine atom can destroy as many as 100,000 ozone molecules.

There is a lot of controversy about ozone depletion. Some believe that the chlorine in the stratosphere comes from natural sources. The increase in the amount of chlorine in random air samples taken from the stratosphere matches the increase in fluorine, which has a different natural source. The increase in the amount of chlorine and fluorine parallels the increase in CFC emissions over the past two decades. Unlike other chlorine compounds and naturally occurring chlorine, the chlorine in CFCs is not water soluble and is not washed out of the atmosphere during rain.

THREE PRIMARY REFRIGERANT TYPES

The three primary types of refrigerants are:

- **CFC** containing chlorine, fluorine, and carbon
- **HCFC** containing hydrogen, chlorine, fluorine, and carbon
- **HFC** containing hydrogen, fluorine, and carbon

OZONE-DEPLETION POTENTIAL

The ODP is a scale that is used to determine the degree that CFCs and HCFCs destroy the ozone. CFCs, such as R-11, R-12, and R-500, have the highest ODP. HCFCs, such as R-22 and R-123, contain lesser amounts of chlorine, and are not as hazardous as CFCs. HFCs, such as R-134a, do not contain chlorine and have no ODP.

THE CLEAN AIR ACT AND THE MONTREAL PROTOCOL

After years of negotiations, an international treaty was formed to regulate the production and use of CFCs, halons, methyl chloroform, carbon tetrachloride, and various other ozone-depleting substances. Known as the

Certification

Montreal Protocol, this landmark agreement in mid-1989 initially called for a production and consumption freeze of some ozone-depleting substances. It currently calls for a step reduction and eventual phaseout of all ozone-depleting substances.

The greatest offender, CFC R-12, was phased out of production on December 31, 1995. When virgin supplies are depleted, all future supplies will come from recovered, recycled, and reclaimed CFCs.

The EPA set standards for refrigerant recovery prior to appliance repair or disposal. To stop damage to the stratospheric ozone layer, the U.S. Congress enacted amendments to the CAA in 1990. The amendments mandated the capturing and eventual phaseout of CFC and HCFC production. It is a violation of Section 608 to:

- Intentionally release CFCs or HCFCs while repairing appliances
- Fail to achieve the required evacuation level before opening or disposing of an appliance
- Fail to have an EPA-approved recovery device equipped with low-loss fittings
- Falsify or fail to keep required records
- Fail to become certified (after November 14, 1994)
- Vent CFCs or HCFCs (after July 1, 1992)
- Vent HFC refrigerants (after November 15, 1995)
- Release mixtures of nitrogen and refrigerant from a fully charged system while testing

In addition to EPA regulations, some state and local government regulations contain rules as strict as or stricter than those of Section 608.

RECOVERY DEVICES

Refrigerant recovery and recycling equipment manufactured after November 15, 1993, must be certified by an EPA-approved equipment testing organization, such as the Underwriters' Laboratory (UL) or the Environmental Technology Laboratory (ETL), to meet ARI standards.

There are basically two types of recovery devices. One type referred to as passive is system dependent and relies on the compressor within the refrigeration appliance to remove the refrigerant. Passive recovery devices may be used only on appliances that contain 15 pounds (6.8 kg) or less of refrigerant.

Larger appliances with nonoperating compressors cannot achieve mandated evacuation rates using system-dependent equipment. The EPA, therefore, requires that technicians who repair or dispose of large appliances have at least one self-contained recovery device available. This is required to recover refrigerant from appliances that have nonoperating compressors.

The other type of equipment is referred to as active. It is self-contained and has its own means of drawing refrigerant out of the appliance. An active recovery device can be used on any size of appliance. Active recovery devices are capable of removing refrigerant from an appliance whether the compressor is operating or not.

RECOVERY TECHNIQUES

There are a few things one should remember about the proper methods of refrigerant recovery. First, before beginning any recovery procedure, be sure to check the nameplate to see what type of refrigerant is in the system. This is important to ensure that refrigerants in a recovery cylinder are not mixed. Each recovery cylinder must be designated for a specific refrigerant type. Mixed refrigerants are generally impossible to reclaim. A reclamation facility may refuse to reclaim or may charge a substantial fee to destroy mixed refrigerants.

The recovery machine should be checked for proper oil level and leaks as part of a (generally daily) schedule. When using recovery equipment or cylinders with Schrader-type valves, inspect the valve cores for leaks. Always cap Schrader ports when not in use to prevent accidental depression of the valve core. EPA-approved recovery equipment must be equipped with low-loss fittings. Low-loss fittings minimize the release of refrigerant when hoses are disconnected. The length and size of the hoses between the appliance and the recovery machine will affect the efficiency of the recovery process. Long or small hoses increase recovery time and the potential for accidental release of refrigerants.

Recovery time can be reduced by recovering as much refrigerant as possible in the liquid phase. Be sure that your recovery equipment is designed to recover liquid. Liquid recovery removes the maximum amount of oil as well as contaminants from the system. This recovery should begin at the lowest point on the system, such as the evaporator liquid line on a roof-mounted condensing unit split system.

After recovering the liquid refrigerant, the remaining vapor is condensed when removed by the recovery unit. Recovery time can also be reduced by heating the appliance and/or cooling the recovery cylinder.

Recovered refrigerant may only be returned to the appliance from which it was removed or to another appliance owned by the same person. If the appliance is a motor vehicle air-conditioning (MVAC) like appliance, the refrigerant must be recycled or reclaimed before it may be reused. Because recovered refrigerants may contain contaminants such as acids and moisture, it is necessary to periodically check and replace the oil and filter on a recycling machine.

When filling a charging cylinder with refrigerant, the refrigerant vapor that is vented off the top of the cylinder must be recovered.

RECOVERY CYLINDERS

Recovery cylinders are designed to be refilled. They differ in many ways from disposable cylinders. Disposable refrigerant cylinders, used with virgin refrigerant, may not be used as a recovery cylinder. To do so is a violation of federal law. Empty or near-empty disposable cylinders must be disposed of by recovering any remaining refrigerant and by rendering the cylinder useless.

Recovery cylinders should not be overfilled or heated. This practice could cause an explosion. The EPA mandates that a refillable refrigerant cylinder must be filled to no more than 80 percent of its capacity by weight. The safe filling level must be controlled by a mechanical float device, an electronic shut-off device, or by weight. Before an empty recovery cylinder may be put into service, it should be evacuated to remove any air and moisture that may be present. Air in a refrigeration system will cause higher than normal discharge pressures. Before using recovered/recycled refrigerant, the recovery cylinder should be allowed to stabilize at room temperature so it can be checked for noncondensables. When checking for the presence of noncondensable gases, comparisons with a temperature/pressure chart can only be made at a stabile, accurately measured ambient temperature.

Reusable cylinders used to transport recovered pressurized refrigerant must be approved by the DOT. An approved refrigerant cylinder should be color coded with a gray body and a yellow top. All recovery cylinders must be hydrostatically tested and date stamped every five years. If a cylinder shows signs of rust or abuse or does not appear to be sound, it should not be used.

SHIPPING AND TRANSPORTING

Before shipping used refrigerant in recovery cylinders, all shipping paperwork must be completed properly. This includes having the correct number of cylinders of each refrigerant, checking that the cylinder meets DOT standards, and properly labeling the cylinder with the type and amount of refrigerant. Cylinders should be transported in an upright position and marked with a DOT diamond tag indicating it is classified as a "2.2 nonflammable gas." Some states may require special shipping procedures to be followed depending on their classification of used refrigerants. Check with the DOT and the EPA offices in the state of origin.

DEHYDRATION

Proper dehydration is a necessity. This process is easily accomplished through standard evacuation procedures. If moisture is allowed to remain in a system that contains CFCs or HCFCs, the refrigerants decompose and form hydrochloric and hydrofluoric acids. Achieving a deep vacuum is the recommended method to remove moisture from a system. Also, pulling a deep vacuum and monitoring the vacuum level is a good method of determining whether the system has a leak.

Do not evacuate a system to the ambient air without first following the proper and prescribed recovery procedures, maintaining the mandated vacuum level.

Some factors that contribute to the speed and efficiency of dehydration by evacuation are ambient temperature, size of the appliance, quantity of moisture contamination, capacity of the vacuum pump, and the length and diameter of the vacuum suction hose. The vacuum hose should be short and as large or larger in diameter than the vacuum pump inlet. Measurement of the vacuum level must be performed while the system is isolated. The vacuum gauge should be as far from the vacuum pump as possible. Bigger is not necessarily better for a vacuum pump. A vacuum pump with a high capacity may lower the pressure too rapidly and cause any moisture in the system to freeze. If the system is contaminated with a large amount of moisture, increasing the pressure with nitrogen will prevent freezing. Dehydration is complete when the vacuum gauge shows that you have reached and held the required maximum vacuum.

LEAK REPAIR REQUIREMENTS

Although it may be a good business practice as well as important for the environment to repair all leaks, the EPA does not require leak repair of small appliances.

Appliances other than commercial refrigeration and industrial process refrigeration that contain more than 50 pounds (22.7 kg) of refrigerant must be repaired if the annual leak rate exceeds 15 percent of the total system capacity.

Commercial refrigeration, which must be repaired when the annual leak rate exceeds 35 percent, includes appliances used in the retail food and cold storage warehouse sectors (including equipment found in supermarkets, convenience stores, restaurants and other food establishments, and equipment used to store meat, produce, dairy products, and other perishable goods.)

Industrial process refrigeration has the same leak repair requirements as commercial refrigeration and includes complex customized appliances used in the chemical, pharmaceutical, petrochemical, and manufacturing industries, as well as industrial ice machines and ice rinks.

SUBSTITUTE REFRIGERANTS AND LUBRICANTS

New refrigerants, blends of older refrigerants, and different lubricants are being constantly introduced. This means that we will have to become familiar with and keep up with new products, terms, and procedures as they develop.

HFC-134a, the leading candidate for CFC-12 retrofit, is not a drop-in substitute. HFC-134a (R-134a) can be used in most CFC-12 (R-12) systems by following proper retrofit procedures. However, R-134a will not mix with mineral-based refrigerant oils. The oils used with most HFCs, such as R-134a, are esters. Esters cannot be mixed with other refrigerant oils. It is also important when leak testing a retrofitted R-134a system that you use only R-134a refrigerant or pressurize with nitrogen so as not to cause an accidental refrigerant mix.

There are several refrigerant blends in use. Some blends are called ternary, meaning that they are comprised of three parts. If the system develops a leak, a ternary blend will leak at different rates because of the different vapor pressures of each component of the blend. The variance in the boiling and condensing temperatures of the ternary refrigerants is called *temperature glide*. The proper method of charging a blend into a system is to weigh the refrigerant into the high side as a liquid. Ternary blends are used with a synthetic alkyl benzene oil. It is important that the correct oil for the refrigerant is used. Most refrigerant oils are hygroscopic, meaning they rapidly absorb moisture. For that reason, always keep oil containers sealed tightly when not in use to prevent moisture contamination.

Another type of blend is known as azeotropic. An azeotropic blend combines on a molecular level to create a refrigerant with its own temperature/pressure characteristics. An azeotropic mixture does not have a temperature glide, so it performs just like any conventional refrigerant.

SAFETY

The EPA is concerned with the prevention of refrigerant venting, and with technicians' overall safety and well-being. Therefore, when handling and filling refrigerant cylinders or operating recovery equipment, technicians should wear safety glasses, protective gloves, and follow all manufacturers' safety precautions. This is most important when handling the new refrigerants where toxicity concerns exist. Always review the material safety data sheets (MSDS) when working with any solvents, chemicals, or refrigerants.

In large quantities, refrigerants can cause suffocation because they are several times heavier than air and displace the oxygen that we breathe. Asphyxia, a term for oxygen deprivation, is the number one cause of death in refrigeration-related accidents. Providing adequate ventilation is a must in low areas where refrigerants may accumulate. Inhaling refrigerant vapors may cause heart irregularities and unconsciousness. If a high concentration of refrigerant vapor develops in a confined area, immediately vacate and ventilate the area. Only persons using, and properly trained in the use of, a self-contained breathing apparatus (SCBA) should enter an enclosed area where a refrigerant release is known to have occurred.

Do not expose R-12 or R-22 to open flames or glowing hot metal surfaces. These, as well as most halogenated compounds, will decompose to form hydrofluoric acid and hydrochloric acid. When there is a source of moisture or oxygen present, phosgene gas may be formed. These acids have a sharp stinging effect on the nose, and, if detected, the area should be vacated and ventilated immediately.

Certification

Never pressurize the system with oxygen or compressed air for leak testing or for any other reason. Oxygen or compressed air, when mixed with refrigerant oil, can cause an explosion. If pressurizing a system with nitrogen, always charge through a pressure regulator. There should be a relief valve downstream from the pressure regulator but not be installed in series. If corrosion is found in the body of a relief valve, it must be replaced. To determine the safe pressure for leak testing, check the data plate for the low-side test pressure value. Many refrigeration systems are also protected with the use of a pressure relief valve.

The American Society of Heating, Refrigeration, and Air Conditioning Engineers (ASHRAE) Standard 15 requires that all equipment rooms containing equipment using an A-l refrigerant, such as R-11, R-12, and R-134a be fitted with an oxygen-deprivation sensor and alarm. Equipment rooms with equipment using B-l refrigerants, such as R-123, must be fitted with a refrigerant sensor.

Here are a few *nevers* to keep in mind:

- Never start a compressor when the discharge valve is closed.
- Never start a hermetic compressor while under a vacuum.
- Never mix refrigerants.
- Never mix refrigerant lubricants.
- Never work in a nonventilated area without adequate protection
- Never allow liquid refrigerant to strike the skin.
- Never allow refrigeration lubricant to strike the skin.

PRACTICE TEST

The following questions are for practice and are similar to those found in the CORE test.

CORE Practice Test Questions

1. The stratospheric ozone depletion is a _____ problem.
    (A) United States
    (B) global
    (C) local and municipal government
    (D) United Nations

2. The condition and state of refrigerant as it enters the condenser of an air-conditioning system is:
   (A) subcooled liquid.
   (B) superheated vapor.
   (C) subcooled vapor.
   (D) superheated liquid.

3. What determines the time required to properly evacuate and dehydrate a system prior to charging with refrigerant?
   (A) The total volume of the system
   (B) The amount on noncondensables in the system
   (C) The ambient temperature surrounding the system
   (D) All of the above.

4. What is the condition and state of the refrigerant as it leaves a receiver?
   (A) Subcooled liquid
   (B) Superheated liquid
   (C) Subcooled vapor
   (D) Superheated vapor

5. How are refrigerant cylinders shipped?
   (A) Upside-down
   (B) Upright
   (C) As a hazardous material
   (D) In crates

6. Which of the following refrigerants contain the greatest amount of chlorine (Cl)?
   (A) HCFCs
   (B) HFCs
   (C) CFCs
   (D) CHFCs

7. When did it become unlawful to intentionally release Class I and Class II refrigerants into the atmosphere?
   (A) January 1, 1992
   (B) July 1, 1992
   (C) January 1, 1994
   (D) July 1, 1994

8. Which of the following refrigerants have the lowest ozone-depletion potential (ODP)?
   (A) CFCs
   (B) HCFCs

(C) HFCs
(D) CHFCs

9. What is the chemical in CFC refrigerant that destroys stratospheric ozone?
   (A) Fluorine
   (B) Carbon
   (C) Hydrogen
   (D) Chlorine

10. Stratospheric ozone consists of:
    (A) one oxygen atom.
    (B) two oxygen atoms.
    (C) three oxygen atoms.
    (D) four oxygen atoms.

11. Which of the following refrigerants is an HFC?
    (A) R-134a
    (B) R-123
    (C) R-22
    (D) R-12

12. What is the component of an air-conditioning system that changes low-pressure vapor to high-pressure vapor?
    (A) Condenser
    (B) Metering device
    (C) Evaporator
    (D) Compressor

13. The gas in the stratosphere that protects us from ultraviolet radiation is:
    (A) radon.
    (B) oxygen.
    (C) nitrogen.
    (D) ozone.

14. Which of the following refrigerants is a CFC?
    (A) R-134a
    (B) R-123
    (C) R-22
    (D) R-12

15. What is the molecule that is found in the stratosphere that indicates ozone depletion is taking place?
    (A) Carbon dioxide
    (B) Carbon monoxide

(C) Chlorine monoxide
(D) Radon

16. Which of the following refrigerants *does not* contain chlorine?
    (A) R-134a
    (B) R-123
    (C) R-22
    (D) R-12

17. Which of the following actions is a violation of the Clean Air Act (CAA)?
    (A) Knowingly releasing Class I and Class II substances
    (B) Failing to keep or falsifying required records
    (C) Failing to reach the required evacuation levels before opening a system to the atmosphere for service
    (D) All of the above.

18. Disposable (one-time use) refrigerant cylinders are used to:
    (A) send used refrigerant in for reclaim.
    (B) containing recovered refrigerant.
    (C) recycling refrigerant.
    (D) packaging virgin refrigerant.

19. Which of the following refrigerants is an HCFC?
    (A) R-134a
    (B) R-123
    (C) R-12
    (D) None of the above.

20. The maximum capacity for refilling approved cylinders is:
    (A) 60 percent.
    (B) 70 percent.
    (C) 80 percent.
    (D) 90 percent.

21. Any system that does not hold a vacuum after the recommended evacuation process:
    (A) May have a restriction.
    (B) May contain moisture.
    (C) Both of the above.
    (D) Neither of the above.

22. To *recover* refrigerant means to:
    (A) remove refrigerant, in any condition, from a system and store it in an external container.
    (B) reduce contaminants in used refrigerant by oil separation through filter driers.

(C) reprocess used refrigerant to new product specifications.
(D) remove refrigerant from a system and change ownership.

23. The process that reduces contaminants in used refrigerant by oil separation with single and multiple passes through devices such as replaceable filter core driers, which reduce moisture, acidity, and particulate matter is:
    (A) restoring.
    (B) recovery.
    (C) recycling.
    (D) reclamation.

24. Synthetic lubricant that is used with HCFC-based ternary refrigerant blends is:
    (A) alkyl benzene.
    (B) PAG.
    (C) mineral.
    (D) ester.

25. R-134a is a replacement refrigerant for:
    (A) R-12.
    (B) R-22.
    (C) R-11.
    (D) R-134a is not a replacement refrigerant.

26. Disposable cylinders are leak tested at:
    (A) 275 psi.
    (B) 300 psi.
    (C) 325 psi.
    (D) 350 psi.

27. Which of the following refrigerants has the highest ozone-depleting potential (ODP)?
    (A) CFC-11
    (B) CFC-12
    (C) CFC-114
    (D) CFC-115

28. Which of the following *is not* a Class I substance?
    (A) CFC-12
    (B) Halon-1211
    (C) Carbon tetrachloride
    (D) HCFC-22

29. A manifold hose may be purged to:
    (A) remove air from the hose.
    (B) remove refrigerant from the hose.
    (C) Either A or B.
    (D) Neither A nor B.

30. What is the color of a CFC-11 disposable refrigerant container?
    (A) Yellow
    (B) Purple
    (C) Silver
    (D) Orange

31. What service pressure rating is required by the DOT for a disposable refrigerant cylinder?
    (A) 230 psi
    (B) 240 psi
    (C) 250 psi
    (D) 260 psi

32. The color of a recovery cylinder with removable cap should be:
    (A) Gray with yellow cap.
    (B) Yellow with gray cap.
    (C) Gray with yellow shoulder and cap.
    (D) Yellow with grey shoulder and cap.

33. What does the following marking on a recovery cylinder mean?
    A2
    01   99
       21
    (A) The date the cylinder was last tested
    (B) The date the cylinder is to be retested
    (C) The date the cylinder was manufactured
    (D) The date the cylinder must be taken out of service

34. What is tare weight?
    (A) The weight of a cylinder with its contents
    (B) The weight of the contents of a cylinder
    (C) The weight of a "full" cylinder
    (D) The weight of an empty cylinder

35. Which of the following *are not* synthetic oils?
    (A) Minerals
    (B) Glycols

(C) Esters
(D) Alkyl benzenes

36. The viscosity rating of lubricant is determined by its:
    (A) pour point.
    (B) floc point.
    (C) SUS rating.
    (D) SAE rating.

37. Which of the following substance must be shipped as a hazardous material?
    (A) Used refrigerant
    (B) Nonflammable gas
    (C) Flammable gas
    (D) Mixed refrigerant

38. Which of the following may result due to exposure to refrigerant?
    (A) Cardiac effects
    (B) Frostbite
    (C) Both A and B
    (D) Neither A nor B

39. Which of the following should be used to raise system pressure for leak testing?
    (A) Oxygen
    (B) Nitrogen
    (C) Compressed air
    (D) Argon

40. The pressure range of a high-side test gauge is:
    (A) 0–300 psi (0–2069 kPa).
    (B) 0–400 psi (0–2758 kPa).
    (C) 0–500 psi (0–3448 kPa).
    (D) 0–600 psi (0–4137 kPa).

41. Which of the following affects the speed of refrigerant recovery?
    (A) The inside diameter (ID) of the recovery hose
    (B) The length of the recovery hose
    (C) The ambient temperature
    (D) All of the above.

42. Which of the following is a low-pressure refrigerant?
    (A) R-12
    (B) R-11

(C) R-22
(D) R-134a

43. Which of the following systems *are not* covered by Section 608?
    (A) MVAC systems
    (B) MVAC-like systems
    (C) CFC systems
    (D) HCFC systems

44. Reclaimed refrigerant should be reprocessed to:
    (A) ARI Standard 700.
    (B) SAE Standard 93-270.
    (C) UL standards.
    (D) EPA standards.

45. The tendency of the components of a blend refrigerant to leak at different rates is called:
    (A) separation.
    (B) segregation.
    (C) fragmentation.
    (D) fractionation.

46. Refrigerant turns to a liquid from a vapor in the:
    (A) compressor.
    (B) condenser.
    (C) receiver.
    (D) evaporator.

47. Refrigerant R-12 is:
    (A) a CFC refrigerant.
    (B) an HCFC refrigerant.
    (C) an HFC refrigerant.
    (D) a CHFC refrigerant.

48. The venting of all refrigerants was banned:
    (A) July 1994.
    (B) August 1993.
    (C) January 1998.
    (D) November 1995.

49. Which of the following is a hydrofluorocarbon (HFC) refrigerant?
    (A) R-12
    (B) R-22
    (C) R-134a
    (D) R-717

50. Which of the following is NOT a synthetic oil?
    (A) Paraffinic oil
    (B) Polyolester oil
    (C) Polyalkylene glycol oil
    (D) Alkyl benzene oil

51. The range of condensing or evaporating temperatures for one pressure is known as the:
    (A) acceptable range.
    (B) latent heat range.
    (C) temperature/pressure relationship.
    (D) temperature glide.

52. In refrigerant nomenclature, the first number to the right indicates the number of:
    (A) fluorine atoms.
    (B) carbon atoms.
    (C) hydrogen atoms.
    (D) chlorine atoms.

53. Ozone is destroyed when a chlorine atom:
    (A) replaces an ozone atom in the stratosphere.
    (B) attacks an ozone molecule, breaking it apart.
    (C) combines with a free oxygen atom.
    (D) unites with other chlorine atoms in the stratosphere.

54. Ozone depletion may have an adverse effect on health by causing:
    (A) skin cancer.
    (B) cataracts.
    (C) Both A and B.
    (D) Neither A nor B.

55. Which of the following contributes most to global warming?
    (A) Sulfur dioxide (SO_2)
    (B) Carbon dioxide (CO_2)
    (C) Carbon monoxide (CO)
    (D) CFC and HCFC refrigerants

56. The United States banned the use of CFCs in nonessential aerosols in:
    (A) 1968.
    (B) 1978.
    (C) 1988.
    (D) 1998.

57. An example of a non-essential aerosol is:
    (A) asthma inhalant spray.
    (B) insect control spray.
    (C) Either A or B.
    (D) Neither A nor B.
58. The Clean Air Act (CAA) Amendment was signed into law by the President of the United States on:
    (A) July 1, 1989.
    (B) November 15, 1990.
    (C) January 1, 1992.
    (D) October 31, 1993.
59. A low-pressure appliance is one that uses refrigerant that boils _____ at atmospheric pressure.
    (A) below -58°F (-50°C).
    (B) between -58°F (-50°C) and 50°F (10°C).
    (C) above 50°F (10°C)
    (D) None of the above.
60. For Universal certification, one must be certified in:
    (A) Type I.
    (B) Type II.
    (C) Type III.
    (D) All of the above.
61. Leak repairs are required of comfort cooling chillers having an annual leak rate of:
    (A) 10 percent.
    (B) 15 percent.
    (C) 20 percent.
    (D) 25 percent.
62. A single violation of regulations. under Title VII, can result in daily fines up to:
    (A) $10,000.
    (B) $15,000.
    (C) $20.000.
    (D) $25,000.
63. Those who report violators under Title VII may get awards up to:
    (A) $10,000.
    (B) $15,000.
    (C) $20,000.
    (D) $25,000.

64. Passive refrigerant recovery requires:
    (A) a functioning compressor.
    (B) a refrigerant recovery system.
    (C) several hours.
    (D) None of the above.

65. What determines the time required for refrigerant recovery from a system?
    (A) Refrigerant type
    (B) Recovery method (liquid or vapor)
    (C) Both A and B
    (D) Neither A nor B

66. A major repair includes the replacement of:
    (A) an evaporator blower motor.
    (B) a condenser coil.
    (C) Both A and B.
    (D) Neither A nor B.

67. A minor repair would be to:
    (A) replace the condenser fan motor.
    (B) add refrigerant to the system.
    (C) Both A and B.
    (D) Neither A nor B.

68. When recovering refrigerant from a small appliance with a nonoperating compressor, how much of the refrigerant must be removed?
    (A) 70 percent
    (B) 80 percent
    (C) 90 percent
    (D) 100 percent

69. A small appliance is one that contains less than ____ pound(s) of refrigerant.
    (A) 1
    (B) 5
    (C) 10
    (D) 50

70. Hard silver solder is best when connecting:
    (A) steel to brass or copper.
    (B) copper to brass.
    (C) Both A and B.
    (D) Neither A nor B.

71. The pressure of a fully charged nitrogen cylinder is about:
    (A) 1000 psig (6895 kPa).
    (B) 1500 psig (10,343 kPa).
    (C) 2000 psig (13,790 kPa).
    (D) 2000 psig (17,238 kPa).
72. There should be a pressure relief valve downstream of a nitrogen regulator at no greater than:
    (A) 100 psig (690 kPa).
    (B) 150 psig (1034 kPa).
    (C) 200 psig (1379 kPa).
    (D) 250 psig (1724 kPa).
73. Which of the following is not true of an HFC refrigerant?
    (A) They contain no chlorine atoms.
    (B) They have low global warming potentials.
    (C) They have shorter atmospheric lives.
    (D) They have low ozone depletion materials.
74. Which of the following is not true of a refrigerant blend?
    (A) A blend can be HCFC based.
    (B) A blend can be HFC based.
    (C) A blend has lower ozone-depletion potential than most CFC refrigerants.
    (D) HFC blends are short-term replacements for CFC and HCFC refrigerants.
75. A glycol lubricant:
    (A) is not compatible with chlorine.
    (B) is very hygroscopic.
    (C) Both A and B.
    (D) Neither A nor B.

TYPE I

SMALL APPLIANCES

Technicians who want to service small appliances must be certified in refrigerant recovery if they perform sealed-system service after November 14, 1995. The EPA definition of a small appliance includes products manufactured, charged, and hermetically sealed in a factory with five pounds (2.268 kg) or less of refrigerant, including packaged terminal air conditioners (PTAC), commonly called "window shakers." Persons handling refrigerant during service, maintenance, or repair of small appliances must be certified technicians in either a Type I or Universal.

AN OVERVIEW OF THE ISSUES ON THE TYPE I TEST

Leak Detection

- ✔ Be able to identify the signs of leakage in high-pressure systems, such as excessive superheat and traces of oil for hermetic systems.
- ✔ Be aware of the need to leak test before charging or recharging equipment.

✔ Know the order of preference for leak test gases. For example, nitrogen alone is best, but nitrogen with a trace quantity of R-22 is better than pure refrigerant.

Leak Repair Requirements

✔ Know the allowable annual leak rate for commercial and industrial process refrigeration as well as the allowable annual leak rate for other appliances containing more than 50 pounds (22.7 kg) of refrigerant.

Recovery Techniques

✔ Be aware of the methods used to speed up the recovery process, such as recovering liquid refrigerant at the beginning of the recovery process. Other methods for speeding recovery include chilling the recovery vessel and heating the appliance or the vessel from which refrigerant is being recovered.

✔ Be aware of the methods used for reducing cross-contamination and emissions when a recovery or recycling machine is used with a new refrigerant.

Recovery Requirements

✔ Know the evacuation requirements for high-pressure appliances in each of the following situations: disposal, major versus nonmajor repairs, leaky versus nonleaky appliances, an appliance or component containing less versus more than 200 pounds (90.7 kg) of refrigerant, and recovery/recycling equipment built before versus after November 15, 1993.

✔ Know the definitions of "major repairs" and "nonmajor repairs." A major repair is one that involves removing the compressor, evaporator, condenser, or other major component.

✔ Know the need to wait a certain period of time, generally a few minutes, after reaching the required recovery vacuum to see whether the system pressure rises, indicating that there is still liquid refrigerant in the system or in the lubricant.

✔ Know the prohibition on using system-dependent recovery equipment on systems containing more than 15 pounds (6.8 kg) of refrigerant.

Refrigeration

✔ Know how to identify the refrigerant in an appliance.
✔ Be able to determine the temperature/pressure relationships of common high-pressure refrigerants using a standard temperature/pressure chart. Be aware of the need to add 14.7 to psig to convert the pressure to psia.
✔ Be able to identify the components of high-pressure appliances, such as the receiver, evaporator, or accumulator.
✔ Be able to identify the state of the refrigerant, vapor, or liquid in each component of a high-pressure appliance.

Safety

✔ Know that a hermetic compressor under a vacuum should not be electrically energized.
✔ Be aware of the equipment room requirements under American Society of Heating, Refrigeration, and Air-Conditioning Engineers (ASHRAE) Standard 15, requiring an oxygen-deprivation sensor with all refrigerants.

EQUIPMENT REQUIREMENTS

After August 12, 1993, anyone using recovery equipment must certify to the EPA that the equipment is capable of removing 80 percent of the refrigerant or achieving a 4-in. Hg vacuum. Failure to register the equipment with the EPA constitutes a violation of the CAA provisions. Depending on the date of manufacture, all EPA-approved equipment, either passive or active, must meet the following standards set by the ARI.

- Equipment that was manufactured before November 15, 1993, must be capable of removing 80 percent of the refrigerant or achieving a 4-in. Hg vacuum.
- Equipment that was manufactured after November 15, 1993, must be capable of removing 90 percent of the refrigerant if the compressor is operating, 80 percent of the refrigerant if the compressor is not operating, or achieving a 4-in. Hg vacuum.

There may be questions about these standards, including the dates, on your test. You should spend a few minutes to memorize this information.

RECOVERY TECHNIQUES

Active recovery equipment is self-contained. This recovery equipment has its own independent means of removing refrigerant from an appliance. It is also capable of reaching the required recovery rates whether the appliance compressor is operating or not.

When using passive or system-dependent recovery equipment, some specific procedures may be required. This, however, depends on the condition of the appliance. A system-dependent recovery device may be used to recover refrigerant and deposit it into a nonpressurized charcoal-lined plastic bag or an activated-charcoal absorption system. These special containers may be used in conjunction with a standard vacuum pump. If the appliance compressor is operable, the recovery system should be connected to an access fitting on the high side of the appliance system. Then operate the compressor to pump the refrigerant out of the appliance. If the appliance compressor is not operational, the recovery equipment should be connected to both the low- and high-side access ports of the appliance. Apply heat to the compressor to help release trapped refrigerant from the compressor oil. If a compressor burnout has occurred, a strong odor will be detected during the recovery process. If this is the case contaminated refrigerant, such as that recovered from a burnout, should be recovered in a separate recovery cylinder. It can then be sent to a reclamation facility to be cleaned. After all the refrigerant has been recovered, nitrogen may be used to flush debris out of the system. The nitrogen may be vented to the atmosphere.

Piercing-type access valves should be used only on copper or aluminum tubing. They should be carefully leak tested after they are installed. If, after installing a piercing-type valve, the system pressure is found to be 0 psig (0 kPa), do not begin the recovery process. There is no refrigerant in the system. Piercing-type service valves have a tendency to leak after a period of time. They are therefore intended for short-term service and should not be left on an appliance as a permanent installation. After the refrigerant charge is recovered, the piercing-type valve should be replaced with a permanent sweat-in service valve.

Any small appliance that contains ammonia, hydrogen, methyl formate, methyl chloride, sulfur dioxide, or water as a refrigerant should not be recovered in current recovery equipment. To do so will damage the equipment and/or contaminate refrigerant in the recovery container.

Certification 61

TYPE I TEST

The following practice questions are typical of those that may be found in the Type I certification exam.

Type I Practice Test Questions

1. The sale of Class I and Class II refrigerants has been restricted to technicians certified by an EPA-approved program since:
   (A) July 1, 1992.
   (B) November 15, 1993.
   (C) August 12, 1993.
   (D) November 14, 1994.

2. The required aperture for service typically found on a small appliance system is:
   (A) a three-way valve.
   (B) a process stub.
   (C) a Schrader valve.
   (D) a gate valve.

3. How are disposable refrigerant cylinders disposed of?
   (A) Prepare them for reuse as compressed air tanks.
   (B) Render them useless and send them to salvage.
   (C) Reuse them as recovery cylinders for refrigerants.
   (D) Refill with refrigerants to be sent for reclamation.

4. Which refrigerant is favored as a replacement for CFC-12 (R-12) in small appliances?
   (A) R-22
   (B) R-502
   (C) R-123
   (D) R-134a

5. For use in maintenance, service, and repair of small appliances, recovery equipment manufactured after _____ must be certified by an EPA-approved organization.
   (A) July 1, 1992
   (B) August 12, 1992
   (C) January 1, 1995
   (D) November 15, 1993

6. Packaged heat pumps with a capacity of less than five pounds (2.27 kg) of refrigerant may be serviced by a technician:
   (A) with Type I certification.
   (B) with Type II certification.
   (C) Either of the above.
   (D) Neither of the above.
7. Which of the following refrigerants has a zero ozone-depletion rating (ODR)?
   (A) CFCs
   (B) HCFCs
   (C) CHFCs
   (D) HFCs
8. Which best defines a Type I small appliance according to the EPA?
   (A) Systems manufactured, charged, and hermetically sealed having a capacity of five pounds (2.27 kg) or less of refrigerant
   (B) Refrigerators, freezers, room air conditioners, and central air conditioners
   (C) Any appliance having a capacity of and charged with more than five pounds (2.27 kg) of refrigerant
   (D) Any appliance charged with less than two pounds (0.91 kg) of refrigerant
9. Recovery equipment for small appliance use manufactured after November 15, 1993, must be capable of recovering:
   (A) 80 percent of the refrigerant when the compressor is not operating or achieve a 4-inch vacuum under ARI 740-1993.
   (B) 90 percent of the refrigerant when the compressor is operating or achieve a 4-inch vacuum under ARI 740-1993.
   (C) Both A and B.
   (D) Neither A nor B.
10. Which of the following refrigerants can be mixed?
    (A) R-12 and R-134a
    (B) R-22 and R-502
    (C) Both A and B.
    (D) Neither A nor B.
11. Vapor from the top of a graduated charging cylinder, when filling, should:
    (A) be vented to the atmosphere, regardless of quantity.
    (B) be vented if quantity does not exceed three pounds (1.36 kg).
    (C) not be vented; it should be added to the system being serviced.
    (D) not be vented; it should be recovered.

12. If the regulations of the Clean Air Act (CAA) change after a technician is certified:
    (A) technicians must retest to retain certification.
    (B) technicians previously certified will be grandfathered.
    (C) it is the technician's responsibility to learn and comply with any future changes in the law.
    (D) certified technicians need not be concerned with future changes.
13. What is the color code for a container of refrigerant R-134a?
    (A) Green
    (B) Light green
    (C) Blue
    (D) Light blue
14. The type of refrigerant most harmful to stratospheric ozone is:
    (A) HCFC.
    (B) CFC.
    (C) HFC.
    (D) CHFC.
15. Which refrigerant contains CFC?
    (A) R-500
    (B) R-22
    (C) R-134a
    (D) R-123
16. HCFC refrigerants are:
    (A) more harmful to stratospheric ozone than CFCs.
    (B) as harmful to stratospheric ozone as CFCs.
    (C) less harmful to stratospheric ozone than CFCs.
    (D) not harmful to stratospheric ozone.
17. Which of the following is an HCFC refrigerant?
    (A) R-11
    (B) R-22
    (C) R-114
    (D) R-717
18. Which of the following refrigerants is chlorine free?
    (A) R-134a
    (B) R-12
    (C) R-22
    (D) R-124
19. Which of the following is an HFC refrigerant?
    (A) R-115
    (B) R-134a

(C) R-22
(D) R-11

20. What chemical in the upper stratosphere is an indication that the ozone layer is being destroyed?
    (A) Carbon monoxide
    (B) Nitrous oxide
    (C) Carbon dioxide
    (D) Chlorine monoxide

21. What is being done in the United States to curtail damage to the stratospheric ozone?
    (A) The use of CFCs is being restricted, and there is a ban on venting.
    (B) Strictly enforcing emission requirements on incinerators and waste treatment facilities.
    (C) Either A or B.
    (D) Neither A nor B.

22. The passive, or system-dependent, recovery process:
    (A) does not require the use of a pump or heat to recover refrigerant.
    (B) uses a pressure relief device to protect the technician and recovery equipment.
    (C) is used to recover refrigerant in nonpressurized containers.
    (D) can only be performed on a system with an operating compressor.

23. Nitrogen tanks should always be equipped with a:
    (A) float sensor.
    (B) regulator.
    (C) red top.
    (D) safety cart.

24. If there is a large release of R-12 or R-22 in a confined area:
    (A) safety goggles and lined butyl gloves are sufficient protection.
    (B) a self-contained breathing apparatus (SCBA) or immediately leaving the area is required.
    (C) respiratory protection is not required.
    (D) dusk or particle masks are required.

25. At high temperatures, such as in an open flame, R-12 and R-22 can decompose to form:
    (A) hydrozine gas.
    (B) helium gas.
    (C) Either A or B.
    (D) Neither A nor B.

Certification

26. When using recovery equipment having Schrader valves, it is necessary to:
    (A) inspect the core for damage. Replace damaged valve cores to prevent leakage.
    (B) cap the Schrader ports to prevent accidental depression of the valve core.
    (C) Either A or B.
    (D) Neither A nor B.

27. The Department of Transportation (DOT) Regulations 49 CFR, require _____ be recorded on the shipping paper for hazard class 2.2, Nonflammable Compressed Gases.
    (A) weight of each cylinder
    (B) total cubic feet of each gas
    (C) total weight of all cylinders
    (D) number of cylinders of each gas

28. A cylinder of recovered R-500, at normal room temperature of about 75°F (23.9°C), will be pressurized to:
    (A) 30 psig (207 kPa).
    (B) 95 psig (655 kPa).
    (C) 75 psig (517 kPa).
    (D) 120 psig (827 kPa).

29. Refrigerants such as sulfur dioxide and methyl chloride used in refrigerators built before ____ should not be recovered with current recovery equipment.
    (A) 1940
    (B) 1950
    (C) 1960
    (D) 1970

30. Which of the following refrigerants is considered as a replacement for R-12 in domestic refrigerators?
    (A) R-141a/141b
    (B) R-14la
    (C) R-22
    (D) R-134a

31. It is recommended that piercing-type access valves be used on the following tubing material(s).
    (A) Copper
    (B) Steel

(C) Both A and B.
(D) Neither A nor B.
32. The following should be performed daily:
    (A) Check vacuum pump and recovery equipment for leaks.
    (B) Check amperage (current draw) of the recovery device.
    (C) Both A and B.
    (D) Neither A nor B.
33. If nitrogen is used to pressurize or blow debris out of the system, the nitrogen:
    (A) must be recovered.
    (B) may be vented.
    (C) should not be used.
    (D) need not be removed.
34. Low-loss fittings are used to:
(A) connect the recovery device to an appliance and can be manually closed or may close automatically when disconnected to prevent loss of refrigerant from hoses.
    (B) connect the recovery device to an appliance and leak only miscible amounts of refrigerant during use.
    (C) connect the recovery device to an appliance and must be discarded after each use.
    (D) reduce refrigerant loss during normal reverse cycle operation.
35. Refrigerant recovery systems must be equipped with _____ fittings.
    (A) low-loss
    (B) quick-connect
    (C) reverse flow
    (D) stop-flow
36. Any technician who receives a passing grade on the small appliance examination is certified to recover refrigerant during the maintenance, service, and repair of:
    (A) small central air-conditioning systems with 10 pounds (4.54 kg) or less of refrigerant.
    (B) packaged terminal air conditioners (PTAC) with 5 pounds or less of refrigerant.
    (C) any low-pressure equipment.
    (D) motor vehicle air-conditioning (MVAC) equipment.
37. Technicians recovering refrigerant during maintenance, service, or repair of small appliances must have:
    (A) Type I certification.
    (B) Type II certification.

(C) Type III certification.
(D) Either of the above.

38. Before starting a refrigerant recovery procedure it is always necessary to:
    (A) allow the appliance to stabilize at room temperature before proceeding.
    (B) move the appliance to an outdoor location.
    (C) know the quantity of refrigerant in the system.
    (D) know the type of refrigerant that is in the system.

39. Using the passive (system-dependent) recovery process, both the high and low side of the system is accessed for refrigerant recovery:
    (A) when there is a leak in the system.
    (B) when the compressor operates normally.
    (C) when there is moisture in the system.
    (D) when the compressor does not run.

40. What is the definition of a small appliance?
    (A) Products manufactured, charged and hermetically sealed in a factory
    (B) Products having five pounds (2.27 kg) or less of refrigerant
    (C) Either A or B.
    (D) Neither A nor B.

41. Recovery systems manufactured before November 15, 1993, for use with small appliances must meet which of the following requirements?
    (A) Must be capable of recovering 80 percent of the refrigerant whether the compressor is operating or achieving a 4-inch vacuum under conditions of ARI 740-1993.
    (B) Must be capable of recovering 70 percent of the refrigerant if the compressor is operating or achieving a 4-inch vacuum under conditions of ARI 740-1993.
    (C) Must be capable of achieving a 10-inch vacuum under conditions of ARI 740-1993.
    (D) There are no special requirements. This equipment would be considered "grandfathered."

42. Recovery equipment for small appliance service, manufactured after November 15, 1993, must be certified to be capable of:
    (A) recovering 90 percent of the refrigerant when the compressor is operating or achieving a 4-inch vacuum under the conditions of ARI 740-1993.

(B) recovering 80 percent of the refrigerant when the compressor is operating or achieving a 4-inch vacuum under the conditions of ARI 740-1993.
(C) recovering 95 percent of the refrigerant when the compressor is operating or achieving a 10-inch vacuum under the conditions of ARI 740-1993.
(D) recovering 75 percent of the refrigerant when the compressor is operating or achieving a 4-inch vacuum under the conditions of ARI 740-1993.

43. Recovery equipment must be certified by an EPA-approved laboratory if manufactured after:
    (A) July 1, 1992.
    (B) May 13, 1993.
    (C) July 1, 1993.
    (D) November 15, 1993.

44. After August 12, 1993, technicians using recovery equipment must certify to the EPA that they have:
    (A) equipment capable of removing 90 percent of the refrigerant when the compressor is operating or achieving 10-inch vacuum under ARI 740-1993.
    (B) equipment capable of removing 80 percent of the refrigerant or achieving a 4-inch vacuum under ARI 740-1993.
    (C) equipment capable of achieving a 27-inch vacuum under conditions of ARI 740-1993.
    (D) Either of the above.

45. Technicians who service small appliances must be certified in:
    (A) refrigerant recovery.
    (B) first aid procedures.
    (C) Either of the above.
    (D) Neither of the above.

46. Refrigerant entering the metering device of a refrigeration system is a:
    (A) saturated vapor.
    (B) liquid.
    (C) superheated vapor.
    (D) mixture of vapor and liquid.

47. Which term describes the process used to return refrigerant to new-product specifications requiring chemical analysis for verification?
    (A) Recovery
    (B) Recycle

(C) Reclaim
(D) Reprocess

48. Which of the following leak detection methods is considered most effective for locating the area of a small leak?
(A) Bubble test
(B) Electronic/ultrasonic test
(C) Halide torch
(D) Audible sound

49. When servicing a system, it is discovered that R-502 was added to an R-22 system. The technician should:
(A) vent the refrigerant since it cannot be reclaimed.
(B) recycle the refrigerant.
(C) recover and use in another system.
(D) recover the refrigerant into a separate tank.

50. Evacuating a system is a method of:
(A) checking for leaks.
(B) removing moisture.
(C) Either A or B.
(D) Neither A nor B.

51. How is the filling level controlled when transferring refrigerant to a pressurized cylinder?
(A) The safe filling level can be controlled by a mechanical float device.
(B) The safe filling level can be controlled by an electronic shut-off device.
(C) The safe filling level can be controlled by weight.
(D) All of the above.

52. Disposable cylinders are used for:
(A) recycled refrigerant.
(B) recovered refrigerant.
(C) contaminated refrigerant.
(D) virgin refrigerant.

53. A refrigerant label is placed on a:
(A) cylinder to be returned for reclaiming.
(B) truck to identify cylinder hauler.
(C) cylinder to identify gross weight.
(D) cylinder to identify pressure.

54. The Clean Air Act (CAA):
(A) calls for the phaseout of CFC/HCFC production.
(B) requires the EPA to set standards for the recovery of refrigerants prior to appliance disposal.

(C) Either A or B.
(D) Neither A nor B.

55. Violation of the Clean Air Act (CAA), including the intentional release of refrigerant during the maintenance, service, repair, or disposal of appliances, can result in:
    (A) a warning without a fine.
    (B) a fine of $1000 per day per violation.
    (C) a fine of $10,000 per day per violation.
    (D) a fine of $25,000 per day per violation.

56. When servicing a refrigeration system containing R-12, the refrigerant must be:
    (A) replaced with R-134a.
    (B) recovered.
    (C) vented.
    (D) destroyed.

57. It has been illegal to vent substitute refrigerants since:
    (A) July 1992.
    (B) November 1993.
    (C) July 1993.
    (D) November 1995.

58. Under EPA's regulations, what standard must reclaimed refrigerant meet before it can be resold?
    (A) ARI 700
    (B) EPA 700
    (C) ASHRAE 700
    (D) UL 700

59. A system-dependent recovery device is one that:
    (A) is hermetically sealed and requires special maintenance.
    (B) must be connected to the liquid port on large systems.
    (C) captures refrigerant with the help of components in the refrigeration equipment.
    (D) must be plugged into a power source.

60. A refrigerant lubricant that is hygroscopic:
    (A) is a poor lubricant.
    (B) is crude oil and unrefined.
    (C) can be left in an open container.
    (D) has a high affinity for water.

Certification

61. What synthetic lubricant is presently being used with ternary blends containing HCFCs?
    (A) alkyl benzene
    (B) ester
    (C) glycol
    (D) mineral
62. What is the state of the refrigerant leaving the condenser?
    (A) low-pressure liquid
    (B) low-pressure vapor
    (C) high-pressure vapor
    (D) high-pressure liquid
63. System-dependent (passive) refrigerant recovery of small appliances:
    (A) requires that 80 percent of the refrigerant be recovered.
    (B) recovers refrigerant in a nonpressurized container.
    (C) Either A or B.
    (D) Neither A nor B.
64. A container of mixed refrigerant sent to a reclamation center will be:
    (A) disposed of at the owner's expense.
    (B) reclaimed and returned.
    (C) reused in fire extinguishers.
    (D) returned to the owner "as is."
65. Before disposing of a small appliance containing R-12, it is necessary to:
    (A) pressurize the system with nitrogen.
    (B) recover the refrigerant.
    (C) make sure that it is not operational.
    (D) dismantle the system.
66. Hydrochloric and hydrofluoric acids are:
    (A) caused by CFCs or HCFCs in the presence of moisture and high temperature.
    (B) damaging to the windings in a hermetic compressor.
    (C) Both A and B.
    (D) Neither A nor B.
67. A refrigeration system has a restriction at the capillary tube inlet. What access is needed to recover the refrigerant?
    (A) One access valve on the high side of the system
    (B) One access valve on the low side of the system
    (C) Two access valves, high and low side of the system
    (D) Two access valves, on the evaporator and the low side

68. The law requires that leaks in small appliances:
   (A) must be repaired.
   (B) must be repaired if refrigerant capacity is more than five pounds (2.27 kg).
   (C) does not have to be repaired unless the leak rate exceeds two pounds (0.9 kg) per year.
   (D) does not have to be repaired.
69. Why can CFC or HCFC refrigerants in the air cause suffocation?
   (A) Refrigerants contain an acidic substance.
   (B) Refrigerants heavier than air displace oxygen.
   (C) Refrigerants lighter than air will rise.
   (D) Refrigerants obnoxious odor prevents breathing.
70. A nitrogen tank should always be equipped with a(n):
   (A) pressure regulator.
   (B) pressure relief valve.
   (C) low-loss fitting.
   (D) air-purge valve.
71. Nitrogen, when used for leak checking a refrigeration system:
   (A) must be recovered.
   (B) will introduce moisture into the system.
   (C) may be vented to the atmosphere.
   (D) need not be removed from the system.
72. Why should line piercing and tapping valves be removed from the system after repair?
   (A) They restrict the flow of refrigerant.
   (B) They are expensive.
   (C) Their gaskets, in time, will fail and cause a leak.
   (D) They destroy the appearance.
73. What is the last year that CFCs were manufactured in the United States?
   (A) 1995
   (B) 1996
   (C) 1997
   (D) 1998
74. What certification is required for a technician who works on small appliances?
   (A) Type I
   (B) Universal

Certification

(C) Either A or B.
(D) Neither A nor B.

75. It is not necessary to recover refrigerants found in absorption systems such as:
    (A) ammonia.
    (B) hydrogen.
    (C) Either A or B.
    (D) Neither A nor B.

TYPE II

HIGH-PRESSURE REFRIGERANT CERTIFICATION

Technicians who want to service, maintain, repair, or dispose of high or very high pressure appliances must be certified as either a Type II or Universal technician. The exceptions are those technicians who service small appliances or motor vehicle air conditioning systems (MVACs).

After November 14, 1994, the sale of CFC and HCFC refrigerants has been restricted to technicians properly certified in refrigerant recovery.

OVERVIEW OF ISSUES ON THE TYPE II TEST

The following is an overview of the material that may be on the high-pressure refrigerant certification test.

Leak Detection

✔ Be able to identify the signs of leakage in high-pressure systems, such as excessive superheat and traces of oil for hermetic systems.
✔ Be aware of the need to leak test before charging or recharging equipment.

Certification

✔ Know the order of preference for leak test gases. For example, nitrogen alone is best, but nitrogen with a trace quantity of R-22 is better than pure refrigerant.

Leak Repair Requirements

✔ Know the allowable annual leak rate for commercial and industrial process refrigeration as well as the allowable annual leak rate for other appliances containing more than 50 pounds (22.7 kg) of refrigerant.

Recovery Techniques

✔ Be aware of the methods used to speed up the recovery process, such as recovering liquid refrigerant at beginning of the recovery process. Other methods for speeding recovery include chilling the recovery vessel or heating the appliance or the vessel from which refrigerant is being recovered.

✔ Be aware of the methods used for reducing cross-contamination and emissions when a recovery or recycling machine is used with a new refrigerant.

Recovery Requirements

✔ Know the evacuation requirements for high-pressure appliances in each of the following situations: disposal, major versus nonmajor repairs, leaky versus nonleaky appliances, an appliance or component containing less versus more than 200 pounds (91 kg) of refrigerant, and recovery/recycling equipment built before versus built after November 15, 1993.

✔ Know the definitions of "major repairs" and "nonmajor" repairs." A major repair is one that involves removing the compressor, evaporator, condenser, or other major component.

✔ Know the need to wait a certain period of time, generally a few minutes, after reaching the required recovery vacuum to see whether the system pressure rises, indicating that there is still liquid refrigerant in the system or in the lubricant.

✔ Know the prohibition on using system-dependent recovery equipment on systems containing more than 15 pounds (6.8 kilograms) of refrigerant.

Refrigeration

- ✔ Know how to identify the refrigerant in an appliance.
- ✔ Be able to determine the pressure-temperature relationships of common high-pressure refrigerants using a standard temperature-pressure chart. Be aware to add 14.7 to pounds per square inch–gauge (psig) to convert the pressure to pounds per square inch–absolute (psia). Conversely, subtracting 14.7 from pounds per square inch–absolute (psia) converts the pressure to pounds per square inch–gauge (psig).
- ✔ Be able to identify the components of high-pressure appliances, such as receiver, evaporator, or accumulator.
- ✔ Be able to identify the state of the refrigerant, vapor, or liquid in each component of a high-pressure appliance.

Safety

- ✔ Know that a hermetic compressor under a vacuum should not be electrically energized.
- ✔ Be aware of the equipment room requirements under American Society of Heating, Refrigeration, and Air-Conditioning Engineers (ASHRAE) Standard 15, requiring an oxygen deprivation sensor with all refrigerants.

LEAK DETECTION

After the installation of any type of system, it should first be pressurized with an inert gas, such as nitrogen, and leak tested. A trace amount of refrigerant, preferably R-22, may be added to help identify leaks. To determine the general area of a leak, an electronic or ultrasonic leak detector may be used. After locating the general area of the leak, use a soap solution to pinpoint the leak.

A refrigeration system having an open drive compressor that has not been used for a period of time is likely to leak at the shaft seal area. A visual inspection of any type of system may show traces of oil. These are generally good indicators of a refrigerant leak. Excessive superheat is also an excellent indicator that there is a leak somewhere in a high-pressure system.

RECHARGING

The proper method of charging a system that contains a large amount of refrigerant is as follows. Please note that any time that a refrigeration system has been opened for service the filter/drier should be replaced.

After the refrigerant recovery and system evacuation are completed, the initial charging should begin in the vapor phase. Breaking a deep vacuum with liquid refrigerant can result in a temperature low enough to cause freezing. Before charging with liquid R-12, for example, a system must be at a saturation temperature of 36°F (2.2°C) which is about 33 psig (227.5 kPa). After reaching the correct saturation temperature, the system may be liquid charged.

RECOVERY

The use of appropriate recovery equipment that has been certified by an EPA-approved testing laboratory, such as UL or ETL, to meet ARI standards is required. The use of nonapproved equipment is in violation of the CAA.

A technician who is working with multiple CFC and HCFC refrigerants must purge the recover/recycle equipment before recovering/recycling a different type of refrigerant. This is accomplished by recovering as much of the original refrigerant as possible. After that the filter must be changed and the recovery/recycle system must be evacuated. A technician working with an HFC such as R-134a must use a different recover/recycle machine, gauge and hose set, vacuum pump, recovery cylinder, and oil containers.

RECOVERY REQUIREMENTS

The table on the following page (Table 2) lists the required recovery levels (in inches of mercury) for Type II appliances.

After reaching the mandated minimum recovery vacuum, wait five minutes to see whether the system pressure rises. System pressure may increase as refrigerant is released from the compressor oil. If the system pressure increases, the recovery procedure must be repeated. Whenever leaks in the system make evacuation to the minimum prescribed level impossible, the system may be evacuated to 0 psig (0 kPa).

TABLE 2: Required Levels of Evacuation for Appliances

	Inches of Mercury Vacuum* Using Equipment Manufactured:	
Type of Appliance	Before Nov. 15, 1993	On or after Nov. 15, 1993
EXCEPT FOR SMALL APPLIANCES, MVACS, AND MVAC-LIKE APPLIANCES		
HCFC-22 appliance** normally containing less than 200 pounds of refrigerant	0	0
HCFC-22 appliance** normally containing 200 pounds or more of refrigerant	4	10
Other high-pressure appliance** normally containing less than 200 pounds of refrigerant (CFC-12, -500, -502, -114)	4	10
Other high-pressure appliance** normally containing 200 pounds or more of refrigerant (CFC-12, -500, -502, -114)	4	15
Very High Pressure Appliance (CFC-13, -503)	0	0
Low-Pressure Appliance (CFC-11, HCFC-123)	25	25 mm Hg absolute

*Relative to standard atmospheric pressure of 29.9" Hg.
**Or isolated component of such an appliance

A parallel compressor system must be isolated to recover the system refrigerant. Failure to do so will result in an open equalization connection that will prevent refrigerant recovery. If the compressor is operational and the system has a receiver, as much refrigerant as possible should be recovered in the system receiver.

The EPA regulations' definition of a "major repair" is any maintenance or service involving the removal of any of the major components, such as compressor, condenser, evaporator, or auxiliary heat exchanger coil.

TYPE II TEST

The following practice questions are similar to those that may be found in the Type II certification test.

Type II Practice Test Questions

1. What is one of the easiest ways to identify the type of refrigerant that is in a system?
   (A) Check the stem of the TXV.
   (B) Check the nameplate data.
   (C) Check the pressure cutout settings.
   (D) Smell the refrigerant.

2. By front-seating a three-way suction service valve, the
   (A) suction inlet to the valve will be closed.
   (B) gauge port of the valve will be closed.
   (C) gauge port and suction inlet will be closed.
   (D) valve port to compressor inlet will be closed.

3. At room ambient temperature of 80°F (26.7°C), what is the approximate pressure of a cylinder containing R-22?
   (A) 143 psig (986 kPa)
   (B) 74 psig (510 kPa)
   (C) 114 psig (786 kPa)
   (D) 212 psig (1462 kPa)

4. Which refrigerant may be used as a trace gas, pressurized with nitrogen, for leak detection?
   (A) R-11
   (B) R-12
   (C) R-22
   (D) R-115

5. Visual oil traces around the inlet fitting of a sight glass may be the indication of:
   (A) a leak at the point of the oil traces.
   (B) excessive oil in the system.
   (C) an overcharge of refrigerant.
   (D) a restriction in the system.

6. What may cause excessive superheat at the evaporator outlet of an air-conditioning system?
   (A) High head pressure
   (B) A dirty condenser
   (C) Insufficient air flow
   (D) Low refrigerant charge

7. What is the maximum leak rate allowed for comfort cooling chillers and all other equipment to require repair under EPA regulations?
   (A) 15 percent
   (B) 25 percent
   (C) 35 percent
   (D) 45 percent
8. The required level of evacuation for recovery equipment manufactured before November 15, 1993, is less than that required for equipment manufactured after that date. What is the early requirement when used on a system with an R-22 charge of less than 200 pounds (91 kg)?
   (A) 0 inches Hg
   (B) 4 inches Hg
   (C) 10 inches Hg
   (D) 15 inches Hg
9. What is the required level of evacuation for recovery equipment manufactured after November 15, 1993, on a system that contains less than 200 pounds (91 kg) of R-22 refrigerant?
   (A) 0 inches Hg
   (B) 4 inches Hg
   (C) 10 inches Hg
   (D) 15 inches Hg
10. What is the required level of evacuation for recovery equipment manufactured after November 15, 1993, on a system containing less than 200 pounds (91 kg) of R-12 refrigerant?
    (A) 0 inches Hg
    (B) 4 inches Hg
    (C) 10 inches Hg
    (D) 15 inches Hg
11. What is the maximum annual leak rate for industrial process and commercial refrigeration equipment required to be repaired under EPA regulations?
    (A) 15 percent
    (B) 25 percent
    (C) 35 percent
    (D) 45 percent
12. Which of the following replacements is defined as a major repair by the EPA?
    (A) Compressor
    (B) Evaporator

(C) Either A or B.
(D) Neither A nor B.

13. By EPA rules, a Type II classification applies to:
    (A) Small appliances containing five pounds (2.27 kg) or less of refrigerant.
    (B) Refrigerators, freezers, and vending machines.
    (C) Low-pressure systems.
    (D) Split air-conditioning systems with five pounds (2.27 kg) or more of refrigerant.

14. Most liquid refrigerants to be recovered from a system will be found in the:
    (A) condenser.
    (B) receiver.
    (C) hoses and lines.
    (D) evaporator.

15. It is the owner's responsibility to maintain records of any and all refrigerant that is added to a system that contains _____ pounds (_ kg) or more of refrigerant.
    (A) 15 (6.8)
    (B) 20 (9.07)
    (C) 35 (15.88)
    (D) 50 (22.68)

16. System-dependent recovery equipment is limited to appliances containing less than ___ pounds (_ kg) of refrigerant.
    (A) 15 (6.8)
    (B) 25 (11.34)
    (C) 50 (22.68)
    (D) 75 (34.02)

17. There are exceptions to the required evacuation levels for recovery equipment that require an appliance be evacuated to only 0 psig. These apply to appliances that:
    (A) are being scrapped for salvage.
    (B) contain a substance that would damage the recovery equipment.
    (C) have a defective compressor.
    (D) have a water-cooled condenser.

18. What is the rule for maximum liquid filling of refrigerant cylinders?
    (A) 60 percent
    (B) 70 percent
    (C) 80 percent
    (D) 90 percent

19. What refrigerant would you expect to find in a split air-conditioning system manufactured before 1995?
    (A) R-11
    (B) R-12
    (C) R-22
    (D) R-502
20. How is time required for recovery of refrigerant shortened?
    (A) Recovering as much liquid as possible in the initial stages
    (B) Cooling the recovery tank in ice water
    (C) Either A or B.
    (D) Neither A nor B.
21. What is the condition and state of refrigerant entering the receiver?
    (A) Superheated low-pressure vapor
    (B) Superheated high-pressure vapor
    (C) Subcooled low-pressure liquid
    (D) Subcooled high-pressure liquid
22. In what units is a deep vacuum, for evacuation and dehydration, measured?
    (A) Atmospheres (psi)
    (B) psig (kPa)
    (C) psia (kPa-abs)
    (D) Microns (inches)
23. Any noncondensables in a refrigeration system will:
    (A) be purged with a recovery machine.
    (B) cause an increase in the high-side pressure.
    (C) Either A or B.
    (D) Neither A nor B.
24. R-134a is a drop in replacement refrigerant for:
    (A) R-12.
    (B) R-22.
    (C) Either A or B.
    (D) Neither A nor B.
25. What is the condition and state of refrigerant leaving the accumulator?
    (A) Superheated high-pressure vapor
    (B) Superheated low-pressure vapor
    (C) Subcooled high-pressure liquid
    (D) Subcooled low-pressure liquid

Certification

26. Which of the following will speed up the recovery process?
    (A) Chilling the recovery vessel.
    (B) Heating the appliance.
    (C) Both A and B.
    (D) Neither A nor B.

27. Recovering liquid before vapor refrigerant:
    (A) prevents removing moisture from the system.
    (B) prevents removing lubricant from the system.
    (C) aids in removing noncondensables.
    (D) speeds up refrigerant recovery time.

28. A vacuum pump has been operated for 15 minutes and then turned off. The low-side gauge then slowly rises to 0 psig. The problem may be:
    (A) a leak in the system.
    (B) a leak in the manifold and gauge set.
    (C) Both A and B.
    (D) Neither A nor B.

29. The vacuum pump has been operated for a half hour and then turned off. The low-side gauge rises 5 in. Hg in five minutes. The problem may be:
    (A) air in the system.
    (B) refrigerant in the system.
    (C) impurities in the system.
    (D) oil in the system.

30. It is important to leak test before charging a system to:
    (A) remove excess air and moisture.
    (B) ensure a proper charge of lubricant.
    (C) Both A and B.
    (D) Neither A nor B.

31. It is important to evacuate a system before charging to:
    (A) remove air and moisture.
    (B) determine the proper charge of refrigerant.
    (C) Both A and B.
    (D) Neither A nor B.

32. The best method for pressurizing a system for leak testing is to use:
    (A) nitrogen.
    (B) nitrogen and R-22.
    (C) R-22.
    (D) compressed air.

33. A major repair would be the replacement of a(n):
    (A) evaporator blower motor.
    (B) evaporator coil assembly.
    (C) Both A and B.
    (D) Neither A nor B.
34. To convert psig to psia, add:
    (A) 14.7 to psia.
    (B) 14.7 to psig.
    (C) 17.4 to psia.
    (D) 17.4 to psig.
35. The state of the refrigerant in a receiver is:
    (A) high-pressure liquid.
    (B) high-pressure vapor.
    (C) low-pressure liquid.
    (D) low-pressure vapor.
36. A refrigerant leak may be determined by:
    (A) excessive superheat.
    (B) oil residue.
    (C) Both A and B.
    (D) Neither A nor B.
37. The same manifold and gauge set may be used for:
    (A) HFC and HCFC service.
    (B) CFC and HCFC service.
    (C) CFC and HFC service.
    (D) None of the above.
38. When recovering refrigerant from a leaking system, what level of vacuum must be reached?
    (A) 5 in. Hg
    (B) 10 in. Hg
    (C) 5 psig
    (D) 0 psig
39. If the ambient temperature is 83°F (28°C) and the refrigerant pressure in an inoperative system is 145 psig (1000 kPa), the system is probably charged with:
    (A) R-22.
    (B) R-12.
    (C) R-11
    (D) R-500

40. A white refrigerant disposable cylinder may:
    (A) be used for the temporary recovery of R-12.
    (B) be used for the temporary recovery of R-22.
    (C) be used as a compressed-air storage tank.
    (D) not be reused for any purpose.

41. What is the grace period for reinspection of a recovery cylinder?
    (A) There is no grace period.
    (B) 30 days
    (C) 60 days
    (D) 90 days

42. A hermetic system is under a vacuum and is to be charged with refrigerant. Which of the following *should not* be done?
    (A) Release the vacuum before admitting refrigerant vapor into the system.
    (B) Start the compressor to draw refrigerant vapor into the system.
    (C) Both A and B.
    (D) Neither A nor B.

43. Equipment room oxygen deprivation sensors are required under:
    (A) EPA standards.
    (B) ASHRAE standards.
    (C) SAE standards.
    (D) ARI standards.

44. The termination date for the production of Class I substances is January 1 of:
    (A) 2000.
    (B) 2002.
    (C) 2004.
    (D) 2005.

45. The termination date for the production of Class II substances is January 1 of:
    (A) 2000.
    (B) 2010.
    (C) 2020.
    (D) 2030.

46. What type certification is required of a technician who wants to service high-pressure refrigeration systems?
    (A) Type I
    (B) Type II

(C) Type III
(D) Universal

47. All alternate refrigerants must be approved for use by:
    (A) the EPA.
    (B) SAE.
    (C) the UL.
    (D) OSHA.

48. State laws may be enacted that:
    (A) enhance federal requirements.
    (B) support federal requirements.
    (C) Both A and B.
    (D) Neither A nor B.

49. Those who want to service air-conditioning and refrigeration equipment must:
    (A) become certified for a particular type of service through an EPA-approved program.
    (B) acquire refrigerant recovery and recycle equipment that complies with the rule.
    (C) establish safe disposal requirements of equipment that enters the waste stream.
    (D) All of the above.

50. What type of system may a Universal certificate holder *not* service, under the 608 Rule?
    (A) Centrifugal coolers
    (B) Small appliances
    (C) Non-MVAC
    (D) MVAC

51. What "charge" may be released to the atmosphere?
    (A) Nitrogen and R-22 used as a holding charge.
    (B) Nitrogen and R-22 used as a leak test gas.
    (C) Both A and B.
    (D) Neither A nor B.

52. When connecting or disconnecting a hose from a system, the refrigerant in the hose:
    (A) must be recovered.
    (B) may be released to the atmosphere.
    (C) must be retained in the hose.
    (D) Either of the above.

Certification

53. How do technicians ensure that they are evacuating a system to the EPA-prescribed percentage level?
    (A) Follow equipment manufacturer's directions.
    (B) Rely on past experience for a best-guess judgment.
    (C) Evacuate for a minimum of a half hour.
    (D) None of the above.

54. Recovered refrigerant may be:
    (A) returned to the same system from which it was removed.
    (B) installed in any other system on the premises.
    (C) sold on the open market to any certified technician.
    (D) processed and then released to the atmosphere.

55. Who has the EPA approved for certifying recovery and recycle equipment?
    (A) UL
    (B) ARI
    (C) Both A and B.
    (D) Neither A nor B.

56. Recovery and recycle equipment built before November 15, 1993:
    (A) may not be used.
    (B) may be "grandfathered."
    (C) must be scrapped out.
    (D) does not meet EPA requirements.

57. A particular appliance must be repaired if its leak rate exceeds 15 percent per year. It has been necessary to add eight pounds (36.3 kg) of refrigerant to a 100-pound (45.4 kg) system after two months. What is the annual leak rate?
    (A) 4.8 percent
    (B) 48 percent
    (C) 8 percent
    (D) 80 percent

58. It has been necessary to add 2.75 pounds (1.25 kg) of refrigerant annually to a system that contains 25 pounds (11.34 kg) of refrigerant. What is the annual leak rate?
    (A) 15 percent
    (B) 13 percent
    (C) 11 percent
    (D) 9 percent

59. An HCFC system containing 18 pounds (8.2 kg) of refrigerant requires 1 pound (0.45 kg) of refrigerant during every quarterly preventive maintenance period. This machine:
    (A) should not be repaired.
    (B) need not be, but should be, repaired.
    (C) must be repaired.
    (D) Should be serviced every month.

60. A system having an excessive leak must be repaired within _____ days of discovery.
    (A) 10
    (B) 30
    (C) 60
    (D) 90

61. A MVAC-like air conditioner may be found in:
    (A) a truck.
    (B) stationary refrigeration applications.
    (C) refrigerated vending machines.
    (D) farm equipment.

62. The sale of refrigerant is restricted to:
    (A) a properly licensed business.
    (B) a properly certified technician.
    (C) Both A and B.
    (D) Neither A nor B.

63. A technician with Type II certification *may not* service:
    (A) high-pressure and very high pressure systems.
    (B) small appliances.
    (C) Both A and B.
    (D) Neither A nor B.

64. A technician with Type I certification *may not* service:
    (A) small appliances.
    (B) MVAC-type appliances.
    (C) Both A and B.
    (D) Neither A nor B.

65. Which of the following procedures may a noncertified technician perform?
    (A) Connect and/or disconnect manifold and gauge hoses to a system
    (B) Add refrigerant to or remove refrigerant from a system

(C) Both A and B.
(D) Neither A nor B.

66. Which of the following should *not* be used for leak testing?
    (A) Compressed air
    (B) Oxygen
    (C) Carbon dioxide
    (D) All of the above.

67. The refrigerant recovery process may be faster by:
    (A) recovering liquid and then vapor.
    (B) recovering vapor before the liquid.
    (C) recovering the vapor and liquid at the same time.
    (D) first evacuating the recovery cylinder.

68. The part of a refrigeration system that removes heat is:
    (A) a condenser.
    (B) the evaporator.
    (C) an absorber.
    (D) a receiver.

69. Another term for heat exchanger may be:
    (A) evaporator
    (B) condenser.
    (C) Both A and B.
    (D) Neither A nor B.

70. At what point in the system should the refrigerant be all liquid?
    (A) When entering the compressor
    (B) When entering the metering device.
    (C) When leaving the compressor
    (D) When leaving the metering device

71. At what point in the system should the refrigerant be all vapor?
    (A) When entering the compressor
    (B) When entering the metering device
    (C) When leaving the compressor
    (D) When leaving the metering device

72. Which type leak detector *is not* used with an HFC system?
    (A) Electronic
    (B) Soap solution
    (C) Dye trace
    (D) Propane gas

73. A low refrigerant charge may cause:
    (A) high discharge pressure.
    (B) low superheat.
    (C) Both A and B.
    (D) Neither A nor B.

74. Low suction pressure may be caused by:
    (A) a restriction.
    (B) a low refrigerant charge.
    (C) a defective evaporator fan.
    (D) Either of the above.

75. High head pressure may be caused by:
    (A) a restriction.
    (B) a low refrigerant charge.
    (C) a defective evaporator fan.
    (D) None of the above.

TYPE III

LOW-PRESSURE REFRIGERANT CERTIFICATION

Those technicians who want to service, maintain, repair, or dispose of low-pressure appliances must be certified either as a Type III technician or as a Universal technician.

OVERVIEW OF ISSUES ON THE TYPE III TEST

The following is an overview of the material that may be on the low-pressure refrigerant certification test.

Leak Detection

✔ Know the order of preference of leak test pressurization methods for low-pressure systems: preferred, hot water method or built-in system heating/pressurization device such as Prevac; second preference, nitrogen and so on.

✔ Recognize the signs of leakage into a low-pressure system such as excessive purging.

✔ Know the maximum leak test pressure for low-pressure centrifugal chillers.

Leak Repair Requirements

✔ Know the allowable annual leak rate for commercial as well as industrial process refrigeration equipment. Also know the allowable annual leak rate for other appliances that contain more than 50 pounds (22.68 kg) of refrigerant:

Recovery Techniques

✔ Know that recovering liquid refrigerant at the beginning of a recovery process helps speed up the process.
✔ Be aware of the need to recover refrigerant vapor in addition to refrigerant liquid.
✔ Know the need to heat oil to 130°F (54°C) before removing it to minimize refrigerant release.
✔ Know the need to circulate or remove the water from a chiller during refrigerant evacuation to prevent freezing.
✔ Know the high-pressure cut-out level of recovery devices used with low-pressure appliances.

Recharging Techniques

✔ Know the need to introduce refrigerant vapor before refrigerant liquid to prevent freezing of the water in the tubes and the need to charge centrifugal systems through the evaporator charging valve.

Recovery Requirements

✔ Know the evacuation requirements for low-pressure appliances in each of the following situations: disposal, major versus nonmajor repairs, leaky versus nonleaky appliances, appliance or component containing less versus more than 200 pounds (91 kg), recovery/recycling equipment built before versus after November 15, 1993.
✔ Know the definition of "major repair" and "nonmajor repair." Major repair is the replacement of a major component, such as the compressor, evaporator, or condenser.

Certification

- ✔ Know the allowable methods for pressurizing a low-pressure system for a nonmajor repair by controlled hot water and a system heating/pressurization device such as Prevac.
- ✔ Understand the need to wait a few minutes after reaching the required recovery vacuum to see whether the system pressure rises, indicating that there is still liquid refrigerant in the system or in the oil.

Refrigeration

- ✔ Understand the purpose of the purge unit in low-pressure systems.
- ✔ Understand the pressure-temperature relationships of low-pressure refrigerants.

Safety

- ✔ Be aware of equipment room requirements under the American Society of Heating, Refrigeration and Air-Conditioning Engineers (ASHRAE) Standard 15, Oxygen Deprivation Sensor with All Refrigerants. According to this standard, there must be an equipment-room sensor for refrigerant R-123.

LEAK DETECTION

A low-pressure system operates in a vacuum, below atmospheric pressure. Leaks in the system, such as gaskets or fittings, will cause moisture-laden air to enter the system.

The most effective method of leak checking a low-pressure refrigeration system is to pressurize it. This is best accomplished by the use of controlled hot water or heater blankets. Using nitrogen to pressurize the system is also an acceptable alternate. Controlled hot water can be used to pressurize the unit prior to opening a system to conduct a nonmajor repair.

The EPA definition of a **major repair** is any maintenance, service, or repair requiring the removal of any of the following components: compressor, condenser, evaporator, or auxiliary heat exchanger coil.

Do not exceed 10 psig (69 kPa) when pressurizing a low-pressure system. Exceeding this pressure can cause the rupture disk, connected to the evaporator on a centrifugal chiller, to fail. The rupture disk is a type of nonresetting pressure relief valve that must be replaced if it fails.

A hydrostatic tube test kit is used to leak test a tube. To leak test a water box, remove the water and insert the leak-detector probe in the drain valve.

RECOVERY

A recovery system's high-pressure cut-out is set for 10 psig (69 kPa). A low-pressure recovery vessel's rupture disk will rupture at 15 psig (103 kPa).

Successful recovery of refrigerant from a system using CFC-11 or HCFC-123 starts by removing the liquid and recovering the remaining vapor. To prevent freezing during liquid recovery, the water in the system must be circulated. During vapor recovery the recovery compressor, the system water pumps, and the recovery condenser water must be on. The local water supply is used for the recovery unit condensing coil.

Determine the approximate amount of vapor remaining in an R-11 system at 0 psig (0 kPa). First, assume that there are 3 pounds (1.36 kg) of refrigerant per ton of refrigeration. Next, assume that 10 percent of the refrigerant will remain a vapor after liquid removal. Although this assumption is not exact, the answer will be close, giving a reasonable estimate of the number of pounds of vapor remaining in the system to be recovered.

RECOVERY REQUIREMENTS

The following information should be memorized:

TABLE 3: Required Levels of Evacuation

Using recovery or recycling equipment manufactured or imported before November 15, 1993 manufactured or imported after November 15, 1993	25" hg vacuum 25 mm hg absolute

After the minimum required vacuum has been reached as noted in Table 3, wait for at least 5 minutes and note the system pressure. A pressure rise indicates that there is refrigerant remaining in the system. If this is the case, the recovery procedure must be repeated. ASHRAE guidelines state that during vacuum testing if the system's pressure rises from 1 mm hg to above 2.5 mm hg, the system should be checked for leaks. When system leaks make evacuation to the mandated level impossible, the system should be evacuated to the lowest attainable level prior to a major repair.

The recovery of refrigerant from the oil can be maximized by increasing the temperature to 130°F (54°C).

RECHARGING

After recovery and evacuation are completed, refrigerant charging should begin in the vapor phase. Breaking a deep vacuum with liquid refrigerant can lower the temperatures sufficiently to cause water in the tubes to freeze. Before charging with liquid R-11, the system should be at a saturation temperature of 36°F (2.2°C), or 16.9" hg in the shell.

The refrigerant is introduced into the system through the evaporator charging valve because the evaporator is located below the condenser.

PURGE UNITS

Any leak in a low-pressure system will allow air, a noncondensable, to enter the system. Air or other noncondensables will accumulate at the top of the condenser and must be purged from the system. A purge unit is used to remove air and other noncondensables from the top of the chiller condenser. The purge unit returns the refrigerant removed in the purge process to the system. A purge unit requires frequent leak testing and must be repaired when necessary.

TYPE III TEST

The following questions are typical of those that may be found in a Type III certification exam.

TYPE III Practice Test Questions

1. What do R-11 and R-123 refrigerants have in common?
   (A) They are oxygen depleting.
   (B) They are Class I and II substances as defined by the EPA.

(C) Both A and B.
(D) Neither A nor B.

2. What is the required evacuation level for recovering refrigerant from a low-pressure appliance with a recovery machine manufactured before November 15, 1993?
   (A) 0 psig
   (B) 4 inch Hg
   (C) 10 inch Hg
   (D) 25 inch Hg

3. How is frost best removed from a sight glass?
   (A) Reverse the refrigeration cycle.
   (B) Scrape or chip the ice off.
   (C) Spray the sight glass with alcohol.
   (D) Turn off the water supply.

4. What is the color code for a recovery cylinder for R-22?
   (A) Gray with yellow top
   (B) Yellow with gray top
   (C) Yellow with white top
   (D) White with yellow top

5. Why do low-pressure chillers require purge units?
   (A) They operate below atmospheric pressure.
   (B) They draw noncondensables through gaskets and seals.
   (C) Either A or B.
   (D) Neither A nor B.

6. What is connected to the rupture disk of a low-pressure chiller?
   (A) The low side
   (B) The high side
   (C) The compressor
   (D) The purge valve

7. The purge unit draws from the:
   (A) rupture disk.
   (B) suction of the compressor.
   (C) evaporator.
   (D) top of the condenser.

8. Before a low-pressure appliance is dismantled for salvage, the refrigerant must be:
   (A) recovered.
   (B) recycled.

(C) reclaimed.
(D) vented.

9. Machine-room safety standards are established by:
   (A) ARI-7000.
   (B) ARI-740.
   (C) ASHRAE 34.
   (D) ASHRAE 15.

10. Recovery machines that use water as a condensing medium generally use:
    (A) cooling tower water.
    (B) municipal water.
    (C) condensate water.
    (D) chilled water.

11. What is the color code for a recovery cylinder for R-12?
    (A) Yellow with gray top
    (B) Gray with yellow top
    (C) Yellow with white top
    (D) White with yellow top

12. Type III certification, as classified by the EPA, is for technicians working on:
    (A) small appliances.
    (B) high-pressure appliances.
    (C) very high-pressure appliances.
    (D) low-pressure appliances.

13. What is the maximum pressure that should be applied to a low-pressure chiller when leak checking with nitrogen?
    (A) 3 psig (20.69 kPa)
    (B) 10 psig (68.95 kPa)
    (C) 20 psig (137.9 kPa)
    (D) 30 psig (206.85 kPa)

14. What is the short-term replacement refrigerant for R-11 chillers?
    (A) R-22
    (B) R-123
    (C) R-500
    (D) R-134a

15. What is the required evacuation level for recovering refrigerant from a low-pressure appliance with a recovery machine manufactured after November 15, 1993?
    (A) 25 mm Hg absolute
    (B) 29 inch Hg

(C) Both A and B.
(D) Neither A nor B.

16. At what pressure is the rupture disk on a low-pressure chiller generally set?
    (A) 3 psig (20.69 kPa)
    (B) 8 psig (55.16 kPa)
    (C) 10 psig (68.95 kPa)
    (D) 15 psig (103.43 kPa)

17. To prevent freezing of the water coils of a low-pressure chiller, it is recommended that:
    (A) when charging, begin with vapor phase.
    (B) do not inject liquid during charging until saturation temperature is above 32°F (0°C).
    (C) Either A or B.
    (D) Neither A nor B.

18. A low-pressure chiller, providing comfort cooling, having an annual leak rate of ____% of the total charge, must be repaired.
    (A) 15
    (B) 20
    (C) 35
    (D) 50

19. Under what safety group is R-134a classified under ASHRAE 34 standard?
    (A) A1
    (B) A2
    (C) B1
    (D) B2

20. Why is refrigerant vapor initially charged into a low-pressure chiller?
    (A) To prevent safety shut down
    (B) To prevent water freeze
    (C) Either A or B.
    (D) Neither A nor B.

21. Under what safety group is R-123 classified under ASHRAE 34 standards?
    (A) A1
    (B) A2
    (C) B2
    (D) B1

22. How are water tube leaks in a low-pressure chiller usually found?
    (A) By noting water puddles in the area of the leak
    (B) By seeing frosted coils in the area of the leak

(C) By performing a hydrostatic tube test
(D) By using an appropriate leak detector

23. Under ASHRAE Standard 15, which of the following refrigerants require equipment-room sensors?
    (A) R-12
    (B) R-500
    (C) R-134a
    (D) R-123

24. Under ASHRAE Standard 15, which of the following refrigerants require equipment-room oxygen deprivation sensors?
    (A) R-12
    (B) R-134a
    (C) Either A or B.
    (D) Neither A nor B.

25. What component will fail first if excessive nitrogen pressure is exerted within a low-pressure chiller?
    (A) The evaporator coil
    (B) The rupture disk
    (C) The compressor seals
    (D) The purge tank

26. Which of the following CFC refrigerants has been used in low-pressure appliances?
    (A) R-11
    (B) R-12
    (C) Either A or B.
    (D) Neither A nor B.

27. What is the color code for a recovery cylinder for R-502?
    (A) Yellow with gray top
    (B) Gray with yellow top
    (C) Yellow with white top
    (D) White with yellow top

28. Which refrigerants are flammable?
    (A) HFCs
    (B) HCFCs
    (C) Both A and B.
    (D) Neither A nor B.

29. What are the symptoms of exposure to HFC refrigerants above the time weighted average (TWA) limits?
    (A) Feelings of intoxication/dizziness
    (B) Upset stomach/nausea

(C) Both A and B.
(D) Neither A nor B.

30. How do refrigerant vapors concentrate and become dangerous?
    (A) They tend to collect in low areas, near the floor.
    (B) They displace oxygen, leading to asphyxiation.
    (C) Both A and B.
    (D) Neither A nor B.

31. When brazing or welding pipes on an air-conditioning or refrigeration system in an enclosed area:
    (A) put on a self-contained breathing apparatus (SCBA).
    (B) recover the refrigerant and purge the system with nitrogen.
    (C) Both A and B.
    (D) Neither A nor B.

32. Retrofitting a chiller may also include replacement of the:
    (A) motor field assembly.
    (B) coolant media.
    (C) Both A and B.
    (D) Neither A nor B.

33. A refrigerant that may be used to replace CFC-11 in centrifugal chillers is:
    (A) R-134a.
    (B) HCFC-22.
    (C) R-717.
    (D) HCFC-123.

34. Hydrogen chloride and hydrogen fluoride are:
    (A) produced when certain refrigerants are burned.
    (B) cleaning agents used to clean up an oil spill.
    (C) alternate refrigerants to replace Class I and Class II refrigerants.
    (D) None of the above.

35. Which of the following refrigerants, by class, have the lowest toxicity rating, 400 ppm or less?
    (A) Class A
    (B) Class B
    (C) Class C
    (D) Class D

36. Which of the following refrigerants, by class, have the lowest flammability rating?
    (A) Class 1
    (B) Class 2

(C) Class 3
(D) Class 4

37. Which of the following associations *is not* concerned with the air-conditioning and refrigeration trades?
    (A) ASHRAE
    (B) ASE
    (C) ARI
    (D) SAE

38. Which section of the EPA is concerned with commercial refrigeration?
    (A) 606
    (B) 607
    (C) 608
    (D) 609

39. Refrigerant oils are not hazardous waste unless they are:
    (A) mixed with other types of oils.
    (B) subject to recycling and reclamation.
    (C) Both A and B.
    (D) Neither A nor B.

40. Oil is considered hazardous waste by federal rules if tested and found to contain:
    (A) paraffin.
    (B) residual refrigerant.
    (C) moisture.
    (D) cadmium.

41. The preferred method of leak testing a centrifugal chiller is:
    (A) with the use of nitrogen.
    (B) the hot water method.
    (C) the cold water method.
    (D) with the use of compressed air.

42. Excessive purging is a sign of:
    (A) leakage.
    (B) loss of refrigerant.
    (C) Both A and B.
    (D) Neither A nor B.

43. To minimize refrigerant release to what temperature should the oil be heated before removal?
    (A) 130°F (54.4°C)
    (B) 150°F (65.6°C)

(C) 170°F (77.8°C)
(D) 190°F (87.8°C)

44. To prevent freezing during refrigerant evacuation, the water in the chiller should be:
    (A) removed.
    (B) circulated.
    (C) Both A and B.
    (D) Neither A nor B.

45. To prevent freezing water in the tubes during charging procedures:
    (A) introduce refrigerant vapor before liquid.
    (B) introduce refrigerant liquid before vapor.
    (C) introduce vapor and liquid simultaneously.
    (D) run the purge pump while charging refrigerant.

46. A centrifugal system should be charged with refrigerant through the:
    (A) low-side compressor service valve.
    (B) high-side compressor service valve.
    (C) condenser charging valve.
    (D) evaporator charging valve.

47. What is the definition of a major repair?
    (A) The replacement of a compressor.
    (B) The replacement of a purge pump.
    (C) Both A and B.
    (D) Neither A nor B.

48. How is a low-pressure system pressurized for a nonmajor repair?
    (A) By controlled hot water.
    (B) With the use of Prevac.
    (C) Both A and B.
    (D) Neither A nor B.

49. After reaching the required recovery vacuum, the recovery machine is turned off. After a few minutes the pressure rises. This is an indication that there:
    (A) is liquid refrigerant in the oil.
    (B) may be a leak in the system.
    (C) is liquid refrigerant in the system.
    (D) All of the above.

50. An equipment room oxygen deprivation sensor is required under:
    (A) SAE Standard J5560.
    (B) ASHRAE Standard 15.

(C) ARI Standard 27.
(D) EPA Standard 608.

51. Do not exceed ____ when pressurizing a low-pressure system for leak testing.
    (A) 5 psig (34.5 kPa)
    (B) 10 psig (69 kPa)
    (C) 25 psig (172 kPa)
    (D) 50 psig (345 kPa)

52. Overpressurizing for leak detection may:
    (A) cause the rupture disk to fail.
    (B) cause a leak in the heat exchanger.
    (C) rupture a seal or O-ring.
    (D) damage a pressure control device.

53. A rupture disk is:
    (A) an overpressure safety device.
    (B) a nonresetting pressure relief valve.
    (C) Both A and B.
    (D) Neither A nor B.

54. What instrument is used to leak test a tube?
    (A) A hydrometer tester
    (B) A hydrostatic tube test kit
    (C) A hygrometer tester
    (D) A hydraulic pressure test kit

55. A low-pressure recovery system high-pressure cut-out is set for:
    (A) 5 psig (34.5 kPa).
    (B) 10 psig (69 kPa).
    (C) 25 psig (172 kPa).
    (D) 50 psig (345 kPa).

56. As a rule of thumb, how much refrigerant, per ton, is in a low-pressure system?
    (A) One pound (0.454 kg)
    (B) Two pounds (0.907 kg)
    (C) Three pounds (1.36 kg)
    (D) Four pounds (1.81 kg)

57. The recovery of refrigerant from the oil can be maximized by:
    (A) removing the oil.
    (B) heating the oil.

(C) prolonged evacuation.
(D) All of the above.

58. A purge unit:
   (A) removes air and other noncondensables from the top of the chiller condenser.
   (B) returns the refrigerant drawn out in the purge process to the system.
   (C) Both A and B.
   (D) Neither A nor B.

59. A purge unit:
   (A) is a self-contained hermetic unit and seldom requires service.
   (B) may be used to adjust the refrigerant charge in the system.
   (C) Both A and B.
   (D) Neither A nor B.

60. ASHRAE Standard 15 establishes:
   (A) refrigerant purity standards for use in chillers.
   (B) equipment-room safety standards.
   (C) evacuation-level requirements for refrigerant recovery.
   (D) maximum pressure permitted for system leak testing.

61. What type of certification is required for a technician working on low-pressure appliances?
   (A) Type I
   (B) Type II
   (C) Type III
   (D) Either of the above.

62. Cylinders used for recovered refrigerants must be pressure tested every:
   (A) six months.
   (B) year.
   (C) three years.
   (D) five years.

63. Contaminated refrigerant isf destroyed by:
   (A) slowly venting.
   (B) disposing in designated landfill.
   (C) controlled burning.
   (D) reclaiming.

64. The Clean Air Act (CAA) was established to implement the provisions of the:
   (A) Environmental Protection Agency (EPA).
   (B) Montreal Protocol.

Certification

(C) Geneva Conference on the environment.
  (D) United Nations (UN) refrigerant phaseout plan.

65. By law, an alternate refrigerant *must* reduce the overall risk to:
    (A) human health.
    (B) the environment.
    (C) Both A and B.
    (D) Neither A nor B.

66. A system containing a substance known to be hazardous to public health or the environment must:
    (A) have a warning label with the name of the substance.
    (B) be retrofitted with a safe substance within 180 days of discovery.
    (C) not be repaired when it becomes inoperable.
    (D) All of the above.

67. What may be considered a nonessential use of Class I substances?
    (A) As a propellant for nasal spray.
    (B) As a propellant for an aerosol spray.
    (C) Both A and B.
    (D) Neither A nor B.

68. Technicians having Type I, II, or III certifications may purchase any refrigerant, *except* those:
    (A) packaged in containers smaller than 20 pounds (9 kg).
    (B) intended for motor vehicle air conditioning (MVAC) use.
    (C) Both A and B.
    (D) Neither A nor B.

69. To service MVAC systems, the technician must be certified in:
    (A) Type I.
    (B) Type II.
    (C) Either of the above.
    (D) Neither of the above.

70. To service MVAC-type systems, the technician must be certified in:
    (A) Type I.
    (B) Type II.
    (C) Either of the above.
    (D) Neither of the above.

71. A *de minimis* release of refrigerant is one that the technician:
    (A) deliberately caused.
    (B) accidentally caused.

(C) caused due to neglect.
(D) All of the above.

72. Refrigerant recovered may be, without restriction:
    (A) returned to the same system.
    (B) used in another system owned by the same person.
    (C) Both A and B.
    (D) Neither A nor B.

73. Evacuation requirements do not apply to:
    (A) repairs to leaky equipment.
    (B) non-major repairs.
    (C) Both A and B.
    (D) Neither A nor B.

74. Small releases of refrigerant during the connect and disconnect of manifold and gauge set hoses are minimized by the use of:
    (A) Schrader valves.
    (B) protective caps.
    (C) check valves.
    (D) Low-loss fittings.

75. What does the data stamped on a refillable cylinder, as shown here, indicate?

    B4
    03   99
    12

    (A) Last tested March 1999 by tester B412.
    (B) Last tested December 1999 by tester B403.
    (C) To be retested March 1999.
    (D) To be retested December 1999.

Section 3

APPENDIX

SECTION 3: APPENDIX
TABLE OF CONTENTS

SECTION 608—REFRIGERANT RECYCLING RULE — 110

TITLE VI—STRATOSPHERIC OZONE PROTECTION — 121

601—Definitions — 122

602—Listing of class I and II substances (n/a) — 123

603—Monitoring and reporting requirements — 124

604—Phaseout of production and consumption of class I substances — 126

605—Phaseout of production and consumption of class II substances — 131

606—Accelerated schedule — 133

607—Exchange authority — 134

608—National recycling and emission reduction program — 135

609—Servicing of motor vehicle air conditioners — 137

610—Nonessential products containing chlorofluorocarbons	140
611—Labeling	141
612—Safe alternatives policy	144
613—Federal procurement	146
614—Relationship to other laws	146
615—Authority of Administrator	147
616—Transfers among Parties to the Montreal Protocol	148
617—International cooperation	149
618—Miscellaneous provisions	149

TECHNICIAN CERTIFICATION PROGRAMS 151

GLOSSARY 167

SECTION 608—REFRIGERANT RECYCLING RULE

This appendix provides an overview of the refrigerant recycling requirements of Section 608 of the Clean Air Act (CAA) of 1990, as amended, including final regulations published on May 14, 1993 (58 FR 28660), August 19, 1994 (59 FR 42950), and November 9, 1994 (59 FR 55912). The Rule also describes the prohibition on venting that became effective on July 1, 1992.

Table of Contents

Overview   111
The Prohibition on Venting   111
Regulatory Requirements   112
Service Practice Requirements   112
    Evacuation Requirements   112
    Exceptions to Evacuation Requirements   113
    Reclamation Requirement   113
Table 1: Levels of Evacuation   114
Equipment Certification   114
Equipment Grandfathering   115
Refrigerant Leaks   115
Technician Certification   116
Refrigerant Sales Restrictions   117
Certification By Owners of Recycling and Recovery Equipment   117
Reclaimer Certification   118
MVAC-Like Appliances   118
Safe Disposal Requirements   119
Major Record Keeping Requirements   119
Hazardous Waste Disposal   120
Enforcement   120
Planning and Acting for the Future   120

Appendix

Overview

Under Section 608 of the CAA, the EPA has established regulations that:

- Require service practices that maximize recycling of ozone-depleting compounds (both chlorofluorocarbons [CFCs] and hydrochlorofluorocarbons [HCFCs] and their blends) during the servicing and disposal of air-conditioning and refrigeration equipment.
- Set certification requirements for recycling and recovery equipment, technicians, and reclaimers.
- Restrict the sale of refrigerant to certified technicians.
- Require persons servicing or disposing of air-conditioning and refrigeration equipment to certify to the EPA that they have acquired recycling or recovery equipment and are complying with the requirements of the rule.
- Require the repair of substantial leaks in air-conditioning and refrigeration equipment with a charge of greater than 50 pounds.
- Establish safe disposal requirements to ensure removal of refrigerants from goods that enter the waste stream with the charge intact (e.g., motor vehicle air conditioners, home refrigerators, and room air conditioners).

The Prohibition On Venting

Effective July 1, 1992, Section 608 of the Act prohibits individuals from knowingly venting ozone-depleting compounds (generally CFCs and HCFCs) used as refrigerants into the atmosphere while maintaining, servicing, repairing, or disposing of air-conditioning or refrigeration equipment (appliances). Only four types of releases are permitted under the prohibition:

1. "De minimis" quantities of refrigerant released in the course of making good-faith attempts to recapture and recycle or safely dispose of refrigerant.
2. Refrigerants emitted in the course of normal operation of air-conditioning and refrigeration equipment (as opposed to during the maintenance, servicing, repair, or disposal of this equipment) such as from mechanical purging and leaks. However, the EPA requires the repair of leaks above a certain size in large equipment (see Refrigerant Leaks).

3. Releases of CFCs or HCFCs that are not used as refrigerants. For instance, mixtures of nitrogen and R-22 that are used as holding charges or as leak test gases may be released, because in these cases the ozone-depleting compound is not used as a refrigerant. However, a technician may not avoid recovering refrigerant by adding nitrogen to a charged system; before nitrogen is added, the system must be evacuated to the appropriate level in Table 1. Otherwise, the CFC or HCFC vented along with the nitrogen will be considered a refrigerant. Similarly, pure CFCs or HCFCs released from appliances will be presumed to be refrigerants, and their release will be considered a violation of the prohibition on venting.
4. Small releases of refrigerant that result from purging hoses or from connecting or disconnecting hoses to charge or service appliances will not be considered violations of the prohibition on venting. However, recovery and recycling equipment manufactured after November 15, 1993, must be equipped with low-loss fittings.

Regulatory Requirements

Service Practice Requirements

1. Evacuation Requirements

Since July 13, 1993, technicians have been required to evacuate air-conditioning and refrigeration equipment to established vacuum levels when opening the equipment. If the technician's recovery or recycling equipment was manufactured any time before November 15, 1993, the air-conditioning and refrigeration equipment must be evacuated to the levels described in the first column of Table 1. If the technician's recovery or recycling equipment was manufactured on or after November 15, 1993, the air-conditioning and refrigeration equipment must be evacuated to the levels described in the second column of Table 1, and the recovery or recycling equipment must have been certified by an EPA-approved equipment testing organization. Persons who simply add refrigerant to (top-off) appliances are not required to evacuate the systems.

Technicians repairing small appliances, such as household refrigerators, window air conditioners, and water coolers, must recover 80 percent of the refrigerant when the technician uses recovery or recycling equipment manufactured before November 15, 1993, or the compressor in the appliance is not operating; OR 90 percent of the refrigerant when the technician uses recovery or recycling equipment manufactured after November 15, 1993, and the compressor in the appliance is operating.

In order to ensure that they are recovering the correct percentage of refrigerant, technicians must use the recovery equipment according to the directions of its manufacturer. Technicians may also satisfy recovery requirements by evacuating the small appliance to four inches of mercury vacuum.

2. Exceptions to Evacuation Requirements

The EPA has established limited exceptions to its evacuation requirements for (1) repairs to leaky equipment and (2) repairs that are not major and that are not followed by an evacuation of the equipment to the environment.

If, due to leaks, evacuation to the levels in Table 1 is not attainable, or would substantially contaminate the refrigerant being recovered, persons opening the appliance must: isolate leaking from nonleaking components wherever possible; evacuate nonleaking components to the levels in Table 1; and evacuate leaking components to the lowest level that can be attained without substantially contaminating the refrigerant. This level cannot exceed 0 psig.

If evacuation of the equipment to the environment is not to be performed when repairs are complete, and if the repair is not major, then the appliance must be evacuated to at least 0 psig before it is opened if it is a high- or very high-pressure appliance; or be pressurized to 0 psig before it is opened if it is a low-pressure appliance. Methods that require subsequent purging (e.g., nitrogen) cannot be used except with appliances containing R-113.

3. Reclamation Requirement

The EPA has also established that refrigerant recovered and/or recycled can be returned to the same system or other systems owned by the same person without restriction. If refrigerant changes ownership, however, that refrigerant must be reclaimed (i.e., cleaned to the ARI 700-1993 standard of purity and chemically analyzed to verify that it meets this standard) unless the refrigerant was used only in a motor vehicle air conditioner (MVAC) or MVAC-like appliance and will be used in the same type of appliance. (Refrigerant used in MVACs and MVAC-like appliances is subject to the purity requirements of the MVAC regulations at 40 CFR Part 82 Subpart B.) The EPA updates the list of reclaimers as new companies are added.

TABLE 1: Required Levels of Evacuation
(except for small appliances, MVACs, and MVAC-like appliances)

Type of appliance	Inches of Mercury vacuum[a] Using equipment manufactured:	
	Before 11/15/93	On/After 11/15/93
HCFC-22 appliance[b] normally containing less that 200 pounds of refrigerant	0	0
HCFC-22 appliance[b] normally containing more that 200 pounds of refrigerant	4	10
Other high-pressure appliance[b] normally containing less than 200 pounds of refrigerant (CFC-12, -500, -502, -114)	4	15
Very high-pressure appliance (CFC-13, -503)	0	0
Low-pressure appliance (CFC-11, HCFC-123)	25	25 mm Hg absolute

[a] Relative to standard atmospheric pressure at 29.9 in. Hg.
[b] Or isolated component of such an appliance.

Equipment Certification

The Agency has established a certification program for recovery and recycling equipment. Under the program, the EPA requires that equipment manufactured on or after November 15, 1993, be tested by an EPA-approved testing organization to ensure that it meets EPA requirements. Recycling and recovery equipment intended for use with air-conditioning and refrigeration equipment besides small appliances must be tested under the ARI 740-1993 test protocol. Recovery equipment intended for use with small appliances must be tested under the ARI 740-1993 protocol.

The Agency requires recovery efficiency standards that vary depending on the size and type of air-conditioning or refrigeration equipment being serviced. For recovery and recycling equipment intended for use with air-conditioning and refrigeration equipment besides small appliances, these standards are the same as those in the second column of Table 1. Recovery equipment intended for use with small appliances must be able to recover 90 percent of the refrigerant in the small appliance when the small appliance compressor is operating and 80 percent of the refrigerant in the small appliance when the compressor is not operating.

The EPA has approved both the Air-Conditioning and Refrigeration Institute (ARI) and Underwriters Laboratories (UL) to certify recycling and recovery equipment. Certified equipment can be identified by a label reading: "This equipment has been certified by ARI/UL to meet EPA's minimum requirements for recycling and/or recovery equipment intended for use with [appropriate category of appliance—e.g., small appliances, HCFC appliances containing less than 200 pounds of refrigerant, all high-pressure appliances, etc.]." Lists of certified equipment may be obtained by contacting the ARI at 703-524-8800 and the UL at 708-272-8800, ext. 42371.

Equipment Grandfathering

Equipment manufactured before November 15, 1993, including home-made equipment, may be grandfathered if it meets the standards in the first column of Table 1. Third-party testing is not required for equipment manufactured before November 15, 1993, but equipment manufactured on or after that date, including home-made equipment, must be tested by a third party (Equipment Certification).

Refrigerant Leaks

Owners of equipment with charges of greater than 50 pounds are required to repair leaks in the equipment when those leaks together would result in the loss of more than a certain percentage of the equipment's charge over a year. For the commercial and industrial process refrigeration sectors, leaks must be repaired when the appliance leaks at a rate that would release 35 percent or more of the charge over a year. For all other sectors, including comfort cooling, leaks must be repaired when the appliance leaks at a rate that would release 15 percent or more of the charge over a year.

The trigger for repair requirements is the current leak rate rather than the total quantity of refrigerant lost. For instance, owners of a commercial refrigeration system containing 100 pounds of charge must repair leaks if they find that the system has lost 10 pounds of charge over the past month; although 10 pounds represents only 10 percent of the system charge in this case, a leak rate of 10 pounds per month would result in the release of over 100 percent of the charge over the year. To track leak rates, owners of air-conditioning and refrigeration equipment with more than 50 pounds of charge must keep records of the quantity of refrigerant added to their equipment during servicing and maintenance procedures.

Owners are required to repair leaks within 30 days of discovery. This requirement is waived if, within 30 days of discovery, owners develop a one-year retrofit or retirement plan for the leaking equipment. Owners of industrial process refrigeration equipment may qualify for additional time under certain circumstances. For example, if an industrial process shutdown is required to repair a leak, owners have 120 days to repair the leak. Owners of leaky industrial process refrigeration equipment should see the Compliance Assistance Guidance Document for Leak Repair (available from the hotline) for additional information concerning time extensions and pertinent record keeping and reporting requirements. The EPA anticipates putting this document on the Web site, but does not have an estimated date for when that will happen.

Technician Certification

The EPA has established a technician certification program for persons ("technicians") who perform maintenance, service, repair, or disposal who could be reasonably expected to release refrigerants into the atmosphere. The definition of "technician" specifically includes and excludes certain activities as follows: attaching and detaching hoses and gauges to and from the appliance to measure pressure within the appliance; adding refrigerant to or removing refrigerant from the appliance; any other activity that violates the integrity of the MVAC-like appliances and small appliances.

In addition, apprentices are exempt from certification requirements provided the apprentice is closely and continually supervised by a certified technician.

The Agency has developed four types of certification:

- For servicing small appliances (Type I).
- For servicing or disposing of high- or very high-pressure appliances, except small appliances and MVACs (Type II).
- For servicing or disposing of low-pressure appliances (Type III).
- For servicing all types of equipment (Universal).

Technicians are required to pass an EPA-approved test given by an EPA-approved certifying organization to become certified under the mandatory program. The Stratospheric Ozone Hotline distributes lists of approved testing organizations.

For further information concerning matters relating to stratospheric ozone protection, call the Stratospheric Ozone Hotline at 800-296-1996 between 10:00 a.m. and 4:00 p.m. ET Monday through Friday.

Refrigerant Sales Restrictions

Under Section 609 of the Clean Air Act, sales of CFC-12 in containers smaller than 20 pounds are restricted solely to technicians certified under EPA's motor vehicle air-conditioning regulations. Persons servicing appliances other than motor vehicle air conditioners may still buy containers of CFC-12 larger than 20 pounds.

Effective November 14, 1994, the sale of refrigerant in any sized container is restricted to technicians certified either under the program described in Technician Certification above or under EPA's motor vehicle air conditioning regulations. The sales restriction covers refrigerant contained in bulk containers (cylinders or drums) and precharged parts. The restriction excludes refrigerant contained in refrigerators or air conditioners with fully assembled refrigerant circuits (e.g., household refrigerators, window air conditioners, and packaged air conditioners), pure HFC refrigerants, and CFCs or HCFCs that are not intended for use as refrigerants. In addition, a restriction on sale of precharged split systems has been stayed (suspended) while the EPA reconsiders this provision.

Certification By Owners Of Recycling And Recovery Equipment

The EPA requires that persons servicing or disposing of air-conditioning and refrigeration equipment certify to the appropriate EPA Regional Office that they have acquired (built, bought, or leased) recovery or recycling equipment and that they are complying with the applicable requirements of this rule. This certification must be signed by the owner of the equipment or another responsible officer and sent to the appropriate EPA Regional Office. Although owners of recycling and recovery equipment are required to list the number of trucks based at their shops, they do not need to have a piece of recycling or recovery equipment for every truck. Owners do not have to send in a new form each time they add recycling or recovery equipment to their inventory.

Reclaimer Certification

Reclaimers are required to return refrigerant to the purity level specified in ARI Standard 700-1993 (an industry-set purity standard) and to verify this purity using the laboratory protocol set forth in the same standard. In addition, reclaimers must release no more than 1.5 percent of the refrigerant during the reclamation process and must dispose of wastes properly. Reclaimers must certify to the Section 608 Recycling Program Manager at EPA headquarters that they are complying with these requirements and that the information given is true and correct. Certification must also include the name and address of the reclaimer and a list of equipment used to reprocess and to analyze the refrigerant.

The EPA encourages reclaimers to participate in third-party reclaimer certification programs, such as that operated by the Air-Conditioning and Refrigeration Institute (ARI). Third-party certification can enhance the attractiveness of a reclaimer's product by providing an objective assessment of its purity. The EPA maintains a list of approved reclaimers that is available from the hotline. In addition, a checklist helps prospective reclaimers provide appropriate information for EPA to review.

MVAC-Like Appliances

Some of the air conditioners that are covered by this rule are identical to motor vehicle air conditioners (MVACs), but they are not covered by the MVAC refrigerant recycling rule (40 CFR Part 82, Subpart B) because they are used in vehicles that are not defined as "motor vehicles." These air conditioners include many systems used in construction equipment, farm vehicles, boats, and airplanes. Like MVACs in cars and trucks, these air conditioners typically contain two or three pounds of CFC-12 and use open-drive compressors to cool the passenger compartments of vehicles. (Vehicle air conditioners utilizing HCFC-22 are not included in this group and are therefore subject to the requirements outlined above for HCFC-22 equipment.) The EPA is defining these air conditioners as "MVAC-like appliances" and is applying the MVAC rule's requirements for the certification and use of recycling and recovery equipment to them. That is, technicians servicing MVAC-like appliances must "properly use" recycling or recovery equipment that has been certified to meet the standards in 40

CFR Part 82, Subpart B. In addition, the EPA is allowing technicians who service MVAC-like appliances to be certified by a certification program approved under the MVAC rule, if they wish.

Safe Disposal Requirements

Under EPA's rule, equipment that is typically dismantled on-site before disposal (e.g., retail food refrigeration, central residential air conditioning, chillers, and industrial process refrigeration) has to have the refrigerant recovered in accordance with EPA's requirements for servicing. However, equipment that typically enters the waste stream with the charge intact (e.g., motor vehicle air conditioners, household refrigerators and freezers, and room air conditioners) is subject to special safe disposal requirements.

Under these requirements, the final person in the disposal chain (e.g., a scrap metal recycler or landfill owner) is responsible for ensuring that refrigerant is recovered from equipment before the final disposal of the equipment. However, persons "upstream" can remove the refrigerant and provide documentation of its removal to the final person if this is more cost-effective.

The equipment used to recover refrigerant from appliances prior to their final disposal must meet the same performance standards as equipment used prior to servicing, but it does not need to be tested by a laboratory. This means that self-built equipment is allowed as long as it meets the performance requirements. For MVACs and MVAC-like appliances, the performance requirement is 102 mm of mercury vacuum; for small appliances, the recovery equipment performance requirements are 90 percent efficiency when the appliance compressor is operational and 80 percent efficiency when the appliance compressor is not operational.

Technician certification is not required for individuals removing refrigerant from appliances in the waste stream.

Major Record Keeping Requirements

Technicians servicing appliances that contain 50 or more pounds of refrigerant must provide the owner with an invoice that indicates the amount of refrigerant added to the appliance. Technicians must also keep a copy of their proof of certification at their place of business.

Owners of appliances that contain 50 or more pounds of refrigerant must keep servicing records documenting the date and type of service, as well as the quantity of refrigerant added.

Wholesalers who sell CFC and HCFC refrigerants must retain invoices that indicate the name of the purchaser, the date of sale, and the quantity of refrigerant purchased.

Reclaimers must maintain records of the names and addresses of persons sending them material for reclamation and the quantity of material sent to them for reclamation. This information must be maintained on a transactional basis. Within 30 days of the end of the calendar year, reclaimers must report to the EPA the total quantity of material sent to them that year for reclamation, the mass of refrigerant reclaimed that year, and the mass of waste products generated that year.

Hazardous Waste Disposal

If refrigerants are recycled or reclaimed, they are not considered hazardous under federal law. In addition, used oils contaminated with CFCs are not hazardous on the condition that:

- They are not mixed with other waste.
- They are subjected to CFC recycling or reclamation.
- They are not mixed with used oils from other sources.

Used oils that contain CFCs after the CFC reclamation procedure, however, are subject to specification limits for used oil fuels if these oils are destined for burning. Individuals with questions regarding the proper handling of these materials should contact EPA's RCRA Hotline at 800-424-9346 or 703-920-9810.

Enforcement

The EPA is performing random inspections, responding to tips, and pursuing potential cases against violators. Under the Clean Air Act, The EPA is authorized to assess fines of up to $25,000 per day for any violation of these regulations.

Planning And Acting For The Future

Observing the refrigerant recycling regulations for Section 608 is essential in order to conserve existing stocks of refrigerants, as well as to comply with Clean Air Act requirements. However, owners of equipment that contains CFC refrigerants should look beyond the immediate need to maintain existing equipment in working order. The EPA urges equipment owners to act now and prepare for the phaseout of CFC production and import, scheduled for January 1, 1996. Owners are advised to begin planning for conversion or replacement of existing equipment with equipment that uses alternative refrigerants.

TITLE VI—STRATOSPHERIC OZONE PROTECTION

Table of Contents

Sec. 601. Definitions   122
Sec. 602. Listing of Class I and Class II Substances   123
Sec. 603. Monitoring and Reporting Requirements   124
Sec. 604. Phaseout of Production and Consumption of Class I Substances   126
Sec. 605. Phaseout of Production and Consumption of Class II Substances   131
Sec. 606. Accelerated Schedule
Sec. 607. Exchange Authority
Sec. 608. National Recycling and Emission Reduction Program   135
Sec. 609. Servicing of Motor Vehicle Air Conditioners   137
Sec. 610. Nonessential Products Containing Chlorofluorocarbons   140
Sec. 611. Labeling   141
Sec. 612. Safe Alternatives Policy   144
Sec. 613. Federal Procurement   146
Sec. 614. Relationship to Other Laws   146
Sec. 615. Authority of Administrator   147
Sec. 616. Transfers Among Parties to the Montreal Protocol   148
Sec. 617. International Cooperation   149
Sec. 618. Miscellaneous Provisions   149

SEC. 601. DEFINITIONS

As used in this title -

(1) APPLIANCE. - The term 'appliance' means any device which contains and uses a class I or class II substance as a refrigerant and which is used for household or commercial purposes, including any air conditioner, refrigerator, chiller, or freezer.

(2) BASELINE YEAR. - The term 'baseline year' means -
  (A) the calendar year 1986, in the case of any class I substance listed in Group I or II under section 602(a),
  (B) the calendar year 1989, in the case of any class I substance listed in Group III, IV, or V under section 602(a), and
  (C) a representative calendar year selected by the Administrator, in the case of -
    (i) any substance added to the list of class I substances after the publication of the initial list under section 602(a), and
    (ii) any class II substance.

(3) CLASS I SUBSTANCE. - The term 'class I substance' means each of the substances listed as provided in section 602(a).

(4) CLASS II SUBSTANCE. - The term 'class II substance' means each of the substances listed as provided in section 602(b).

(5) COMMISSIONER. - The term 'Commissioner' means the Commissioner of the Food and Drug Administration.

(6) CONSUMPTION. - The term 'consumption' means, with respect to any substance, the amount of that substance produced in the United States, plus the amount imported, minus the amount exported to Parties to the Montreal Protocol. Such term shall be construed in a manner consistent with the Montreal Protocol.

(7) IMPORT. - The term 'import' means to land on, bring into, or introduce into, or attempt to land on, bring into, or introduce into, any place subject to the jurisdiction of the United States, whether or not such landing, bringing, or introduction constitutes an importation within the meaning of the customs laws of the United States.

(8) MEDICAL DEVICE. - The term 'medical device' means any device (as defined in the Federal Food, Drug, and Cosmetic Act [21 U.S.C. 321]),

diagnostic product, drug (as defined in the Federal Food, Drug, and Cosmetic Act), and drug delivery system -
- (A) if such device, product, drug, or drug delivery system utilizes a class I or class II substance for which no safe and effective alternative has been developed, and where necessary, approved by the Commissioner; and
- (B) if such device, product, drug, or drug delivery system, has, after notice and opportunity for public comment, been approved and determined to be essential by the Commissioner in consultation with the Administrator.

(9) MONTREAL PROTOCOL. - The terms 'Montreal Protocol' and 'the Protocol' mean the Montreal Protocol on Substances that Deplete the Ozone Layer, a protocol to the Vienna Convention for the Protection of the Ozone Layer, including adjustments adopted by Parties thereto and amendments that have entered into force.

(10) OZONE-DEPLETION POTENTIAL. - The term 'ozone-depletion potential' means a factor established by the Administrator to reflect the ozone-depletion potential of a substance, on a mass per kilogram basis, as compared to chlorofluorocarbon-11 (CFC-11). Such factor shall be based upon the substance's atmospheric lifetime, the molecular weight of bromine and chlorine, and the substance's ability to be photolytically disassociated, and upon other factors determined to be an accurate measure of relative ozone-depletion potential.

(11) PRODUCE, PRODUCED, AND PRODUCTION. - The terms 'produce,' 'produced,' and 'production' refer to the manufacture of a substance from any raw material or feedstock chemical, but such terms do not include -
- (A) the manufacture of a substance that is used and entirely consumed (except for trace quantities) in the manufacture of other chemicals, or
- (B) the reuse or recycling of a substance.

SEC. 602. LISTING OF CLASS I AND CLASS II SUBSTANCES

(n/a)

SEC. 603. MONITORING AND REPORTING REQUIREMENTS

(a) REGULATIONS. - Within 270 days after the enactment of the Clean Air Act Amendments of 1990, the Administrator shall amend the regulations of the Administrator in effect on such date regarding monitoring and reporting of class I and class II substances. Such amendments shall conform to the requirements of this section. The amended regulations shall include requirements with respect to the time and manner of monitoring and reporting as required under this section.

(b) PRODUCTION, IMPORT, AND EXPORT LEVEL REPORTS. - On a quarterly, or such other basis (not less than annually) as determined by the Administrator, each person who produced, imported, or exported a class I or class II substance shall file a report with the Administrator setting forth the amount of the substance that such person produced, imported, and exported during the preceding reporting period. Each such report shall be signed and attested by a responsible officer. No such report shall be required from a person after April 1 of the calendar year after such person permanently ceases production, importation, and exportation of the substance and so notifies the Administrator in writing.

(c) BASELINE REPORTS FOR CLASS I SUBSTANCES. - Unless such information has previously been reported to the Administrator, on the date on which the first report under subsection (b) is required to be filed, each person who produced, imported, or exported a class I substance (other than a substance added to the list of class I substances after the publication of the initial list of such substances under this section) shall file a report with the Administrator setting forth the amount of such substance that such person produced, imported, and exported during the baseline year. In the case of a substance added to the list of class I substances after publication of the initial list of such substances under this section, the regulations shall require that each person who produced, imported, or exported such substance shall file a report with the Administrator within 180 days after the date on which such substance is added to the list, setting forth the amount of the substance that such person produced, imported, and exported in the baseline year.

(d) MONITORING AND REPORTS TO CONGRESS. -

(1) The Administrator shall monitor and, not less often than every three years following enactment of the Clean Air Act Amend-

ments of 1990, submit a report to Congress on the production, use and consumption of class I and class II substances. Such report shall include data on domestic production, use, and consumption, and an estimate of worldwide production, use, and consumption of such substances. Not less frequently than every six years the Administrator shall report to Congress on the environmental and economic effects of any stratospheric ozone depletion.

(2) The Administrators of the National Aeronautics and Space Administration and the National Oceanic and Atmospheric Administration shall monitor, and not less often than every three years following enactment of the Clean Air Act Amendments of 1990, submit a report to Congress on the current average tropospheric concentration of chlorine and bromine and on the level of stratospheric ozone depletion. Such reports shall include updated projections of -

(A) peak chlorine loading;

(B) the rate at which the atmospheric abundance of chlorine is projected to decrease after the year 2000; and

(C) the date by which the atmospheric abundance of chlorine is projected to return to a level of two parts per billion.

Such updated projections shall be made on the basis of current international and domestic controls on substances covered by this title as well as on the basis of such controls supplemented by a year 2000 global 1 phaseout of all halocarbon emissions (the base case). It is the purpose of the Congress through the provisions of this section to monitor closely the production and consumption of class II substances to assure that the production and consumption of such substances will not:

(i) increase significantly the peak chlorine loading that is projected to occur under the base case established for purposes of this section;

(ii) reduce significantly the rate at which the atmospheric abundance of chlorine is projected to decrease under the base case; or

(iii) delay the date by which the average atmospheric concentration of chlorine is projected under the base case to return to a level of two parts per billion.

(e) TECHNOLOGY STATUS REPORT IN 2015. - The Administrator shall review, on a periodic basis, the progress being made in the development of alternative systems or products necessary to manufacture and operate appliances without class II substances. If the Administrator finds, after notice and opportunity for public comment, that as a result of technological development problems, the development of such alternative systems or products will not occur within the time necessary to provide for the manufacture of such equipment without such substances prior to the applicable deadlines under section 605, the Administrator shall, not later than January 1, 2015, so inform the Congress.

(f) EMERGENCY REPORT. - If, in consultation with the Administrators of the National Aeronautics and Space Administration and the National Oceanic and Atmospheric Administration, and after notice and opportunity for public comment, the Administrator determines that the global production, consumption, and use of class II substances are projected to contribute to an atmospheric chlorine loading in excess of the base case projections by more than \5/10\ths parts per billion, the Administrator shall so inform the Congress immediately. The determination referred to in the preceding sentence shall be based on the monitoring under subsection (d) and updated not less often than every 3 years.

SEC. 604. PHASEOUT OF PRODUCTION AND CONSUMPTION OF CLASS I SUBSTANCES

(a) PRODUCTION PHASEOUT. - Effective on January 1 of each year specified in Table 2, it shall be unlawful for any person to produce any class I substance in an annual quantity greater than the relevant percentage specified in Table 2. The percentages in Table 2 refer to a maximum allowable production as a percentage of the quantity of the substance produced by the person concerned in the baseline year.

(b) TERMINATION OF PRODUCTION OF CLASS I SUBSTANCES. - Effective January 1, 2000 (January 1, 2002 in the case of methyl chloroform), it shall be unlawful for any person to produce any amount of a class I substance.

(c) REGULATIONS REGARDING PRODUCTION AND CONSUMPTION OF CLASS I SUBSTANCES. - The Administrator shall promulgate

Appendix

Date	Carbon Tetracholoride	Methyl Chloroform	Other Class I Substances
1991	100%	100%	85%
1992	90%	100%	80%
1993	80%	90%	75%
1994	70%	85%	65%
1995	15%	70%	50%
1996	15%	50%	40%
1997	15%	50%	15%
1998	15%	50%	15%
1999	15%	50%	15%
2000	20%		
2001		20%	

Table 2

regulations within 10 months after the enactment of the Clean Air Act Amendments of 1990 phasing out the production of class I substances in accordance with this section and other applicable provisions of this title. The Administrator shall also promulgate regulations to insure that the consumption of class I substances in the United States is phased out and terminated in accordance with the same schedule (subject to the same exceptions and other provisions) as is applicable to the phaseout and termination of production of class I substances under this title.

(d) EXCEPTIONS FOR ESSENTIAL USES OF METHYL CHLOROFORM, MEDICAL DEVICES, AND AVIATION SAFETY. -

(1) ESSENTIAL USES OF METHYL CHLOROFORM. - Notwithstanding the termination of production required by subsection (b), during the period beginning on January 1, 2002, and ending on January 1, 2005, the Administrator, after notice and opportunity for public comment, may, to the extent such action is consistent with the Montreal Protocol, authorize the production of limited quantities of methyl chloroform solely for use in essential applications (such as nondestructive testing for metal fatigue and corrosion of existing airplane engines and airplane parts susceptible to metal fatigue) for which no safe and effective substitute is available. Notwithstanding this paragraph, the authority to pro-

duce methyl chloroform for use in medical devices shall be provided in accordance with paragraph (2).

(2) MEDICAL DEVICES. - Notwithstanding the termination of production required by subsection (b), the Administrator, after notice and opportunity for public comment, shall, to the extent such action is consistent with the Montreal Protocol, authorize the production of limited quantities of class I substances solely for use in medical devices if such authorization is determined by the Commissioner, in consultation with the Administrator, to be necessary for use in medical devices.

(3) AVIATION SAFETY. -

(A) Notwithstanding the termination of production required by subsection (b), the Administrator, after notice and opportunity for public comment, may, to the extent such action is consistent with the Montreal Protocol, authorize the production of limited quantities of halon-1211 (bromochlorodifluoromethane), halon-1301 (bromotrifluoromethane), and halon-2402 (dibromotetrafluoromethane) solely for purposes of aviation safety if the Administrator of the Federal Aviation Administration, in consultation with the Administrator, determines that no safe and effective substitute has been developed and that such authorization is necessary for aviation safety purposes.

(B) The Administrator of the Federal Aviation Administration shall, in consultation with the Administrator, examine whether safe and effective substitutes for methyl chloroform or alternative techniques will be available for nondestructive testing for metal fatigue and corrosion of existing airplane engines and airplane parts susceptible to metal fatigue and whether an exception for such uses of methyl chloroform under this paragraph will be necessary for purposes of airline safety after January 1, 2005, and provide a report to Congress in 1998.

(4) CAP ON CERTAIN EXCEPTIONS. - Under no circumstances may the authority set forth in paragraphs (1), (2), and (3) of subsection (d) be applied to authorize any person to produce a class I substance in annual quantities greater than 10 percent of that produced by such person during the baseline year.

(e) DEVELOPING COUNTRIES. -

(1) EXCEPTION. - Notwithstanding the phaseout and termination of production required under subsections (a) and (b), the Administrator, after notice and opportunity for public comment, may, consistent with the Montreal Protocol, authorize the production of limited quantities of a class I substance in excess of the amounts otherwise allowable under subsection (a) or (b), or both, solely for export to, and use in, developing countries that are Parties to the Montreal Protocol and are operating under article 5 of such Protocol. Any production authorized under this paragraph shall be solely for purposes of satisfying the basic domestic needs of such countries.

(2) CAP ON EXCEPTION. -

(A) Under no circumstances may the authority set forth in paragraph (1) be applied to authorize any person to produce a class I substance in any year for which a production percentage is specified in Table 2 of subsection (a) in an annual quantity greater than the specified percentage, plus an amount equal to 10 percent of the amount produced by such person in the baseline year.

B) Under no circumstances may the authority set forth in paragraph (1) be applied to authorize any person to produce a class I substance in the applicable termination year referred to in subsection (b), or in any year thereafter, in an annual quantity greater than 15 percent of the baseline quantity of such substance produced by such person.

(C) An exception authorized under this subsection shall terminate no later than January 1, 2010 (2012 in the case of methyl chloroform).

(f) NATIONAL SECURITY. - The President may, to the extent such action is consistent with the Montreal Protocol, issue such orders regarding production and use of CFC-114 (chlorofluorocarbon-114), halon-1211, halon-1301, and halon-2402, at any specified site or facility or on any vessel as may be necessary to protect the national security interests of the United States if the President finds that adequate substitutes are not available and that the production and use of such substance are necessary to protect such national security interest. Such orders may include, where necessary to protect such interests, an exemption from any prohibition or requirement contained in this title. The President shall notify the Congress within 30 days of the issuance of an order under this paragraph providing for any such exemption. Such notification shall include a statement of the reasons for the granting of the exemption. An exemption under this paragraph shall be for a specified period which may not exceed one year. Additional exemptions may be granted, each upon the President's issuance of a new order under this paragraph. Each such additional exemption shall be for a specified period which may not exceed one year. No exemption shall be granted under this paragraph due to lack of appropriation unless the President shall have specifically requested such appropriation as a part of the budgetary process and the Congress shall have failed to make available such requested appropriation.

(g) FIRE SUPPRESSION AND EXPLOSION PREVENTION. -

(1) Notwithstanding the production phaseout set forth in subsection (a), the Administrator, after notice and opportunity for public comment, may, to the extent such action is consistent with the Montreal Protocol, authorize the production of limited quantities of halon-1211, halon-1301, and halon-2402 in excess of the amount otherwise permitted pursuant to the schedule under subsection (a) solely for purposes of fire suppression or explosion prevention if the Administrator, in consultation with the Administrator of the United States Fire Administration, determines that no safe and effective substitute has been developed and that such authorization is necessary for fire suppression or explosion prevention purposes. The Administrator shall not authorize production under this paragraph for purposes of fire safety or explosion prevention training or testing of fire suppression or explosion prevention equipment. In no event shall the Administrator grant an exception under this paragraph that permits production after December 31, 1999.

(2) The Administrator shall periodically monitor and assess the status of efforts to obtain substitutes for the substances referred to in paragraph (1) for purposes of fire suppression or explosion prevention and the probability of such substitutes being available by December 31, 1999. The Administrator, as part of such assessment, shall consider any relevant assessments under the Montreal Protocol and the actions of the Parties pursuant to Article 2B of the Montreal Protocol in identifying essential uses and in permitting a level of production or consumption that is necessary to

satisfy such uses for which no adequate alternatives are available after December 31, 1999. The Administrator shall report to Congress the results of such assessment in 1994 and again in 1998.

(3) Notwithstanding the termination of production set forth in subsection (b), the Administrator, after notice and opportunity for public comment, may, to the extent consistent with the Montreal Protocol, authorize the production of limited quantities of halon-1211, halon-1301, and halon-2402 in the period after December 31, 1999, and before December 31, 2004, solely for purposes of fire suppression or explosion prevention in association with domestic production of crude oil and natural gas energy supplies on the North Slope of Alaska, if the Administrator, in consultation with the Administrator of the United States Fire Administration, determines that no safe and effective substitute has been developed and that such authorization is necessary for fire suppression and explosion prevention purposes. The Administrator shall not authorize production under the paragraph for purposes of fire safety or explosion prevention training or testing of fire suppression or explosion prevention equipment. In no event shall the Administrator authorize under this paragraph any person to produce any such halon in an amount greater than 3 percent of that produced by such person during the baseline year.

SEC. 605. PHASEOUT OF PRODUCTION AND CONSUMPTION OF CLASS II SUBSTANCES

(a) RESTRICTION OF USE OF CLASS II SUBSTANCES. - Effective January 1, 2015, it shall be unlawful for any person to introduce into interstate commerce or use any class II substance unless such substance -

(1) has been used, recovered, and recycled;
(2) is used and entirely consumed (except for trace quantities) in the production of other chemicals; or (3) is used as a refrigerant in appliances manufactured prior to January 1, 2020. As used in this subsection, the term 'refrigerant' means any class II substance used for heat transfer in a refrigerating system.

(b) PRODUCTION PHASEOUT. -

(1) Effective January 1, 2015, it shall be unlawful for any person to produce any class II substance in an annual quantity greater than the quantity of such substance produced by such person during the baseline year.

(2) Effective January 1, 2030, it shall be unlawful for any person to produce any class II substance.

(c) REGULATIONS REGARDING PRODUCTION AND CONSUMPTION OF CLASS II SUBSTANCES. - By December 31, 1999, the Administrator shall promulgate regulations phasing out the production, and restricting the use, of class II substances in accordance with this section, subject to any acceleration of the phaseout of production under section 606. The Administrator shall also promulgate regulations to insure that the consumption of class II substances in the United States is phased out and terminated in accordance with the same schedule (subject to the same exceptions and other provisions) as is applicable to the phaseout and termination of production of class II substances under this title.

(d) EXCEPTIONS. -

(1) MEDICAL DEVICES. -

(A) IN GENERAL. - Notwithstanding the termination of production required under subsection (b)(2) and the restriction on use referred to in subsection (a), the Administrator, after notice and opportunity for public comment, shall, to the extent such action is consistent with the Montreal Protocol, authorize the production and use of limited quantities of class II substances solely for purposes of use in medical devices if such authorization is determined by the Commissioner, in consultation with the Administrator, to be necessary for use in medical devices.

(B) CAP ON EXCEPTION. - Under no circumstances may the authority set forth in subparagraph (A) be applied to authorize any person to produce a class II substance in annual quantities greater than 10 percent of that produced by such person during the baseline year.

(2) DEVELOPING COUNTRIES. -

(A) IN GENERAL. - Notwithstanding the provisions of subsection (a) or (b), the Administrator, after notice and opportunity for public comment, may authorize the production of limited quantities of a class II substance in excess of the quantities otherwise permitted under such provisions solely for export to and use in developing countries that are Parties to the Montreal Protocol, as determined by the Administrator. Any production authorized under this subsection shall be solely for purposes of satisfying the basic domestic needs of such countries.

(B) CAP ON EXCEPTION. -
  (i) Under no circumstances may the authority set forth in subparagraph (A) be applied to authorize any person to produce a class II substance in any year following the effective date of subsection (b)(1) and before the year 2030 in annual quantities greater than 10 percent of the quantity of such substance produced by such person during the baseline year.
  (ii) Under no circumstances may the authority set forth in subparagraph (A) be applied to authorize any person to produce a class II substance in the year 2030, or any year thereafter, in an annual quantity greater than 15 percent of the quantity of such substance produced by such person during the baseline year.
  (iii) Each exception authorized under this paragraph shall terminate no later than January 1, 2040.

SEC. 606. ACCELERATED SCHEDULE

(a) IN GENERAL. - The Administrator shall promulgate regulations, after notice and opportunity for public comment, which establish a schedule for phasing out the production and consumption of class I and class II substances (or use of class II substances) that is more stringent than set forth in section 604 or 605, or both, if -

  (1) based on an assessment of credible current scientific information (including any assessment under the Montreal Protocol) regarding harmful effects on the stratospheric ozone layer associated with a class I or class II substance, the Administrator determines that such more stringent schedule may be necessary to protect human health and the environment against such effects,

(2) based on the availability of substitutes for listed substances, the Administrator determines that such more stringent schedule is practicable, taking into account technological achievability, safety, and other relevant factors, or

(3) the Montreal Protocol is modified to include a schedule to control or reduce production, consumption, or use of any substance more rapidly than the applicable schedule under this title. In making any determination under paragraphs (1) and (2), the Administrator shall consider the status of the period remaining under the applicable schedule under this title.

(b) PETITION. - Any person may petition the Administrator to promulgate regulations under this section. The Administrator shall grant or deny the petition within 180 days after receipt of any such petition. If the Administrator denies the petition, the Administrator shall publish an explanation of why the petition was denied. If the Administrator grants such petition, such final regulations shall be promulgated within one year. Any petition under this subsection shall include a showing by the petitioner that there are data adequate to support the petition. If the Administrator determines that information is not sufficient to make a determination under this subsection, the Administrator shall use any authority available to the Administrator, under any law administered by the Administrator, to acquire such information.

SEC. 607. EXCHANGE AUTHORITY

(a) TRANSFERS. - The Administrator shall, within 10 months after the enactment of the Clean Air Act Amendments of 1990, promulgate rules under this title providing for the issuance of allowances for the production of class I and II substances in accordance with the requirements of this title and governing the transfer of such allowances. Such rules shall insure that the transactions under the authority of this section will result in greater total reductions in the production in each year of class I and class II substances than would occur in that year in the absence of such transactions.

(b) INTERPOLLUTANT TRANSFERS. -

(1) The rules under this section shall permit a production allowance for a substance for any year to be transferred for a production

allowance for another substance for the same year on an ozone depletion weighted basis.

(2) Allowances for substances in each group of class I substances (as listed pursuant to section 602) may only be transferred for allowances for other substances in the same Group.

(3) The Administrator shall, as appropriate, establish groups of class II substances for trading purposes and assign class II substances to such groups. In the case of class II substances, allowances may only be transferred for allowances for other class II substances that are in the same Group.

(c) TRADES WITH OTHER PERSONS. - The rules under this section shall permit two or more persons to transfer production allowances (including interpollutant transfers which meet the requirements of subsections (a) and (b)) if the transferor of such allowances will be subject, under such rules, to an enforceable and quantifiable reduction in annual production which -

(1) exceeds the reduction otherwise applicable to the transferor under this title,

(2) exceeds the production allowances transferred to the transferee, and

(3) would not have occurred in the absence of such transaction.

(d) CONSUMPTION. - The rules under this section shall also provide for the issuance of consumption allowances in accordance with the requirements of this title and for the trading of such allowances in the same manner as is applicable under this section to the trading of production allowances under this section.

SEC. 608. NATIONAL RECYCLING AND EMISSION REDUCTION PROGRAM

(a) IN GENERAL. -

(1) The Administrator shall, by not later than January 1, 1992, promulgate regulations establishing standards and requirements

regarding the use and disposal of class I substances during the service, repair, or disposal of appliances and industrial process refrigeration. Such standards and requirements shall become effective not later than July 1, 1992.

(2) The Administrator shall, within 4 years after the enactment of the Clean Air Act Amendments of 1990, promulgate regulations establishing standards and requirements regarding use and disposal of class I and II substances not covered by paragraph (1), including the use and disposal of class II substances during service, repair, or disposal of appliances and industrial process refrigeration. Such standards and requirements shall become effective not later than 12 months after promulgation of the regulations.

(3) The regulations under this subsection shall include requirements that -

(A) reduce the use and emission of such substances to the lowest achievable level, and

(B) maximize the recapture and recycling of such substances. Such regulations may include requirements to use alternative substances (including substances which are not class I or class II substances) or to minimize use of class I or class II substances, or to promote the use of safe alternatives pursuant to section 612 or any combination of the foregoing.

(b) SAFE DISPOSAL. - The regulations under subsection (a) shall establish standards and requirements for the safe disposal of class I and II substances. Such regulations shall include each of the following -

(1) Requirements that class I or class II substances contained in bulk in appliances, machines, or other goods shall be removed from each such appliance, machine, or other goods prior to the disposal of such items or their delivery for recycling.

(2) Requirements that any appliance, machine, or other goods containing a class I or class II substance in bulk shall not be manufactured, sold, or distributed in interstate commerce or offered for sale or distribution in interstate commerce unless it is equipped with a servicing aperture or an equally effective design feature which will facilitate the recapture of such substance during service and repair or disposal of such item.

(3) Requirements that any product in which a class I or class II substance is incorporated so as to constitute an inherent element of

such product shall be disposed of in a manner that reduces, to the maximum extent practicable, the release of such substance into the environment. If the Administrator determines that the application of this paragraph to any product would result in producing only insignificant environmental benefits, the Administrator shall include in such regulations an exception for such product.

(c) PROHIBITIONS. -
   (1) Effective July 1, 1992, it shall be unlawful for any person, in the course of maintaining, servicing, repairing, or disposing of an appliance or industrial process refrigeration, to knowingly vent or otherwise knowingly release or dispose of any class I or class II substance used as a refrigerant in such appliance (or industrial process refrigeration) in a manner which permits such substance to enter the environment. De minimis releases associated with good-faith attempts to recapture and recycle or safely dispose of any such substance shall not be subject to the prohibition set forth in the preceding sentence.
   (2) Effective five years after the enactment of the Clean Air Act Amendments of 1990, paragraph (1) shall also apply to the venting, release, or disposal of any substitute substance for a class I or class II substance by any person maintaining, servicing, repairing, or disposing of an appliance or industrial process refrigeration which contains and uses as a refrigerant any such substance, unless the Administrator determines that venting, releasing, or disposing of such substance does not pose a threat to the environment. For purposes of this paragraph, the term 'appliance' includes any device which contains and uses as a refrigerant a substitute substance and which is used for household or commercial purposes, including any air conditioner, refrigerator, chiller, or freezer.

SEC. 609. SERVICING OF MOTOR VEHICLE AIR CONDITIONERS

(a) REGULATIONS. - Within one year after the enactment of the Clean Air Act Amendments of 1990, the Administrator shall promulgate regulations in accordance with this section establishing standards and requirements regarding the servicing of motor vehicle air conditioners.

(b) DEFINITIONS. - As used in this section -

(1) The term 'refrigerant' means any class I or class II substance used in a motor vehicle air conditioner. Effective five years after the enactment of the Clean Air Act Amendments of 1990, the term 'refrigerant' shall also include any substitute substance.

(2) (A) The term 'approved refrigerant recycling equipment' means equipment certified by the Administrator (or an independent standards testing organization approved by the Administrator) to meet the standards established by the Administrator and applicable to equipment for the extraction and reclamation of refrigerant from motor vehicle air conditioners. Such standards shall, at a minimum, be at least as stringent as the standards of the Society of Automotive Engineers in effect as of the date of the enactment of the Clean Air Act Amendments of 1990 and applicable to such equipment (SAE standard J-1990).

(B) Equipment purchased before the proposal of regulations under this section shall be considered certified if it is substantially identical to equipment certified as provided in subparagraph (A).

(3) The term 'properly using' means, with respect to approved refrigerant recycling equipment, using such equipment in conformity with standards established by the Administrator and applicable to the use of such equipment. Such standards shall, at a minimum, be at least as stringent as the standards of the Society of Automotive Engineers in effect as of the date of the enactment of the Clean Air Act Amendments of 1990 and applicable to the use of such equipment (SAE standard J-1989).

(4) The term 'properly trained and certified' means training and certification in the proper use of approved refrigerant recycling equipment for motor vehicle air conditioners in conformity with standards established by the Administrator and applicable to the performance of service on motor vehicle air conditioners. Such standards shall, at a minimum, be at least as stringent as specified, as of the date of the enactment of the Clean Air Act Amendments of 1990, in SAE standard J-1989 under the certification program of the National Institute for Automotive Service Excellence

(ASE) or under a similar program such as the training and certification program of the Mobile Air Conditioning Society (MACS).

(c) SERVICING MOTOR VEHICLE AIR CONDITIONERS. - Effective January 1, 1992, no person repairing or servicing motor vehicles for consideration may perform any service on a motor vehicle air conditioner involving the refrigerant for such air conditioner without properly using approved refrigerant recycling equipment and no such person may perform such service unless such person has been properly trained and certified. The requirements of the previous sentence shall not apply until January 1, 1993, in the case of a person repairing or servicing motor vehicles for consideration at an entity which performed service on fewer than 100 motor vehicle air conditioners during calendar year 1990 and if such person so certifies, pursuant to subsection (d)(2), to the Administrator by January 1, 1992.

(d) CERTIFICATION. -
   (1) Effective two years after the enactment of the Clean Air Act Amendments of 1990, each person performing service on motor vehicle air conditioners for consideration shall certify to the Administrator either -
      (A) that such person has acquired, and is properly using, approved refrigerant recycling equipment in service on motor vehicle air conditioners involving refrigerant and that each individual authorized by such person to perform such service is properly trained and certified; or
      (B) that such person is performing such service at an entity which serviced fewer than 100 motor vehicle air conditioners in 1991.
   (2) Effective January 1, 1993, each person who certified under paragraph (1)(B) shall submit a certification under paragraph (1)(A).
   (3) Each certification under this subsection shall contain the name and address of the person certifying under this subsection and the serial number of each unit of approved recycling equipment acquired by such person and shall be signed and attested by the owner or another responsible officer. Certifications under paragraph (1)(A) may be made by submitting the required information to the Administrator on a standard form provided by the manufacturer of certified refrigerant recycling equipment.

(e) SMALL CONTAINERS OF CLASS I OR CLASS II SUBSTANCES. -

Effective two years after the date of the enactment of the Clean Air Act Amendments of 1990, it shall be unlawful for any person to sell or distribute, or offer for sale or distribution, in interstate commerce to any person (other than a person performing service for consideration on motor vehicle air-conditioning systems in compliance with this section) any class I or class II substance that is suitable for use as a refrigerant in a motor vehicle air-conditioning system and that is in a container which contains less than 20 pounds of such refrigerant.

SEC. 610. NONESSENTIAL PRODUCTS CONTAINING CHLOROFLUOROCARBONS

(a) REGULATIONS. - The Administrator shall promulgate regulations to carry out the requirements of this section within one year after the enactment of the Clean Air Act Amendments of 1990.

(b) NONESSENTIAL PRODUCTS. - The regulations under this section shall identify nonessential products that release class I substances into the environment (including any release occurring during manufacture, use, storage, or disposal) and prohibit any person from selling or distributing any such product, or offering any such product for sale or distribution, in interstate commerce. At a minimum, such prohibition shall apply to -

    (1) chlorofluorocarbon-propelled plastic party streamers and noise horns,

    (2) chlorofluorocarbon-containing cleaning fluids for noncommercial electronic and photographic equipment, and

    (3) other consumer products that are determined by the Administrator -

        (A) to release class I substances into the environment (including any release occurring during manufacture, use, storage, or disposal), and

        (B) to be nonessential. In determining whether a product is nonessential, the Administrator shall consider the purpose or intended use of the product, the technological availability of substitutes for such product and for such class I substance, safety, health, and other relevant factors.

(c) EFFECTIVE DATE. - Effective 24 months after the enactment of the Clean Air Act Amendments of 1990, it shall be unlawful for any person to sell or distribute, or offer for sale or distribution, in interstate commerce any nonessential product to which regulations under subsection (a) implementing subsection (b) are applicable.

(d) OTHER PRODUCTS. -
1. Effective January 1, 1994, it shall be unlawful for any person to sell or distribute, or offer for sale or distribution, in interstate commerce -
   - (A) any aerosol product or other pressurized dispenser which contains a class II substance; or
   - (B) any plastic foam product which contains, or is manufactured with, a class II substance.
2. The Administrator is authorized to grant exceptions from the prohibition under subparagraph (A) of paragraph (1) where -
   - (A) the use of the aerosol product or pressurized dispenser is determined by the Administrator to be essential as a result of flammability or worker safety concerns, and
   - (B) the only available alternative to use of a class II substance is use of a class I substance which legally could be substituted for such class II substance.
3. Subparagraph (B) of paragraph (1) shall not apply to -
   - (A) a foam insulation product, or
   - (B) an integral skin, rigid, or semi-rigid foam utilized to provide for motor vehicle safety in accordance with Federal Motor Vehicle Safety Standards where no adequate substitute substance (other than a class I or class II substance) is practicable for effectively meeting such Standards.

(e) MEDICAL DEVICES. - Nothing in this section shall apply to any medical device as defined in section 601(8).

SEC. 611. LABELING

(a) REGULATIONS. - The Administrator shall promulgate regulations to implement the labeling requirements of this section within 18 months after enactment of the Clean Air Act Amendments of 1990, after notice and opportunity for public comment.

(b) CONTAINERS CONTAINING CLASS I OR CLASS II SUBSTANCES AND PRODUCTS CONTAINING CLASS I SUBSTANCES. - Effective 30 months after the enactment of the Clean Air Act Amendments of 1990, no container in which a class I or class II substance is stored or transported, and no product containing a class I substance, shall be introduced into interstate commerce unless it bears a clearly legible and conspicuous label stating:

> 'Warning: Contains [insert name of substance], a substance which harms public health and environment by destroying ozone in the upper atmosphere.'

(c) PRODUCTS CONTAINING CLASS II SUBSTANCES. -

(1) After 30 months after the enactment of the Clean Air Act Amendments of 1990, and before January 1, 2015, no product containing a class II substance shall be introduced into interstate commerce unless it bears the label referred to in subsection (b) if the Administrator determines, after notice and opportunity for public comment, that there are substitute products or manufacturing processes

   (A) that do not rely on the use of such class II substance,

   (B) that reduce the overall risk to human health and the environment, and

   (C) that are currently or potentially available.

(2) Effective January 1, 2015, the requirements of subsection (b) shall apply to all products containing a class II substance.

(d) PRODUCTS MANUFACTURED WITH CLASS I AND CLASS II SUBSTANCES. -

(1) In the case of a class II substance, after 30 months after the enactment of the Clean Air Act Amendments of 1990, and before January 1, 2015, if the Administrator, after notice and opportunity for public comment, makes the determination referred to in subsection (c) with respect to a product manufactured with a process that uses such class II substance, no such product shall be introduced into interstate commerce unless it bears a clearly legible and conspicuous label stating:

'Warning: Manufactured with [insert name of substance], a substance which harms public health and environment by destroying ozone in the upper atmosphere.'

(2) In the case of a class I substance, effective 30 months after the enactment of the Clean Air Act Amendments of 1990, and before January 1, 2015, the labeling requirements of this subsection shall apply to all products manufactured with a process that uses such class I substance unless the Administrator determines that there are no substitute products or manufacturing processes that

(A) do not rely on the use of such class I substance,

(B) reduce the overall risk to human health and the environment, and

(C) are currently or potentially available.

(e) PETITIONS. -

(1) Any person may, at any time after 18 months after the enactment of the Clean Air Act Amendments of 1990, petition the Administrator to apply the requirements of this section to a product containing a class II substance or a product manufactured with a class I or II substance which is not otherwise subject to such requirements. Within 180 days after receiving such petition, the Administrator shall, pursuant to the criteria set forth in subsection (c), either propose to apply the requirements of this section to such product or publish an explanation of the petition denial. If the Administrator proposes to apply such requirements to such product, the Administrator shall, by rule, render a final determination pursuant to such criteria within 1 year after receiving such petition.

(2) Any petition under this paragraph shall include a showing by the petitioner that there are data on the product adequate to support the petition.

(3) If the Administrator determines that information on the product is not sufficient to make the required determination, the Administrator shall use any authority available to the Administrator under any law administered by the Administrator to acquire such information.

(4) In the case of a product determined by the Administrator, upon petition or on the Administrator's own motion, to be subject to the requirements of this section, the Administrator shall establish

an effective date for such requirements. The effective date shall be 1 year after such determination or 30 months after the enactment of the Clean Air Act Amendments of 1990, whichever is later.

(5) Effective January 1, 2015, the labeling requirements of this subsection shall apply to all products manufactured with a process that uses a class I or class II substance.

(f) RELATIONSHIP TO OTHER LAW. -

(1) The labeling requirements of this section shall not constitute, in whole or part, a defense to liability or a cause for reduction in damages in any suit, whether civil or criminal, brought under any law, whether Federal or State, other than a suit for failure to comply with the labeling requirements of this section.

(2) No other approval of such label by the Administrator under any other law administered by the Administrator shall be required with respect to the labeling requirements of this section.

SEC. 612. SAFE ALTERNATIVES POLICY

(a) POLICY. - To the maximum extent practicable, class I and class II substances shall be replaced by chemicals, product substitutes, or alternative manufacturing processes that reduce overall risks to human health and the environment.

(b) REVIEWS AND REPORTS. - The Administrator shall -

(1) in consultation and coordination with interested members of the public and the heads of relevant Federal agencies and departments, recommend Federal research programs and other activities to assist in identifying alternatives to the use of class I and class II substances as refrigerants, solvents, fire retardants, foam blowing agents, and other commercial applications and in achieving a transition to such alternatives, and, where appropriate, seek to maximize the use of Federal research facilities and resources to assist users of class I and class II substances in identifying and developing alternatives to the use of such substances

as refrigerants, solvents, fire retardants, foam blowing agents, and other commercial applications;

(2) examine in consultation and coordination with the Secretary of Defense and the heads of other relevant Federal agencies and departments, including the General Services Administration, Federal procurement practices with respect to class I and class II substances and recommend measures to promote the transition by the Federal Government, as expeditiously as possible, to the use of safe substitutes;

(3) specify initiatives, including appropriate intergovernmental, international, and commercial information and technology transfers, to promote the development and use of safe substitutes for class I and class II substances, including alternative chemicals, product substitutes, and alternative manufacturing processes; and

(4) maintain a public clearinghouse of alternative chemicals, product substitutes, and alternative manufacturing processes that are available for products and manufacturing processes which use class I and class II substances.

(c) ALTERNATIVES FOR CLASS I OR II SUBSTANCES. - Within two years after enactment of the Clean Air Act Amendments of 1990, the Administrator shall promulgate rules under this section providing that it shall be unlawful to replace any class I or class II substance with any substitute substance which the Administrator determines may present adverse effects to human health or the environment, where the Administrator has identified an alternative to such replacement that -

(1) reduces the overall risk to human health and the environment; and

(2) is currently or potentially available. The Administrator shall publish a list of

(A) the substitutes prohibited under this subsection for specific uses and

(B) the safe alternatives identified under this subsection for specific uses.

(d) RIGHT TO PETITION. - Any person may petition the Administrator to add a substance to the lists under subsection (c) or to remove a substance from either of such lists. The Administrator shall grant or deny the petition within 90 days after receipt of any such petition. If the Administrator denies the petition, the Administrator shall publish an explanation of why the petition was denied. If the Administrator grants such petition the Administrator shall publish such revised list within 6 months thereafter. Any petition under this subsection shall include a showing by the petitioner that there are data on the substance adequate to support the petition. If the Administrator determines that information on the substance is not sufficient to make a determination under this subsection, the Administrator shall use any authority available to the Administrator, under any law administered by the Administrator, to acquire such information.

(e) STUDIES AND NOTIFICATION. - The Administrator shall require any person who produces a chemical substitute for a class I substance to provide the Administrator with such person's unpublished health and safety studies on such substitute and require producers to notify the Administrator not less than 90 days before new or existing chemicals are introduced into interstate commerce for significant new uses as substitutes for a class I substance. This subsection shall be subject to section 114(c).

SEC. 613. FEDERAL PROCUREMENT

Not later than 18 months after the enactment of the Clean Air Act Amendments of 1990, the Administrator, in consultation with the Administrator of the General Services Administration and the Secretary of Defense, shall promulgate regulations requiring each department, agency, and instrumentality of the United States to conform its procurement regulations to the policies and requirements of this title and to maximize the substitution of safe alternatives identified under section 612 for class I and class II substances. Not later than 30 months after the enactment of the Clean Air Act Amendments of 1990, each department, agency, and instrumentality of the United States shall so conform its procurement regulations and certify to the President that its regulations have been modified in accordance with this section.

SEC. 614. RELATIONSHIP TO OTHER LAWS

(a) STATE LAWS. - Notwithstanding section 116, during the two-year period beginning on the enactment of the Clean Air Act Amendments of

1990, no State or local government may enforce any requirement concerning the design of any new or recalled appliance for the purpose of protecting the stratospheric ozone layer.

(b) MONTREAL PROTOCOL. - This title as added by the Clean Air Act Amendments of 1990 shall be construed, interpreted, and applied as a supplement to the terms and conditions of the Montreal Protocol, as provided in Article 2, paragraph 11 thereof, and shall not be construed, interpreted, or applied to abrogate the responsibilities or obligations of the United States to implement fully the provisions of the Montreal Protocol. In the case of conflict between any provision of this title and any provision of the Montreal Protocol, the more stringent provision shall govern. Nothing in this title shall be construed, interpreted, or applied to affect the authority or responsibility of the Administrator to implement Article 4 of the Montreal Protocol with other appropriate agencies.

(c) TECHNOLOGY EXPORT AND OVERSEAS INVESTMENT. - Upon enactment of this title, the President shall -

(1) prohibit the export of technologies used to produce a class I substance;

(2) prohibit direct or indirect investments by any person in facilities designed to produce a class I or class II substance in nations that are not Parties to the Montreal Protocol; and

(3) direct that no agency of the government provide bilateral or multilateral subsidies, aids, credits, guarantees, or insurance programs, for the purpose of producing any class I substance.

SEC. 615. AUTHORITY OF ADMINISTRATOR

If, in the Administrator's judgment, any substance, practice, process, or activity may reasonably be anticipated to affect the stratosphere, especially ozone in the stratosphere, and such effect may reasonably be anticipated to endanger public health or welfare, the Administrator shall promptly promulgate regulations respecting the control of such substance, practice, process, or activity, and shall submit notice of the proposal and promulgation of such regulation to the Congress.

SEC. 616. TRANSFERS AMONG PARTIES TO THE MONTREAL PROTOCOL

(a) IN GENERAL. - Consistent with the Montreal Protocol, the United States may engage in transfers with other Parties to the Protocol under the following conditions:

(1) The United States may transfer production allowances to another Party if, at the time of such transfer, the Administrator establishes revised production limits for the United States such that the aggregate national United States production permitted under the revised production limits equals the lesser of

(A) the maximum production level permitted for the Substance or substances concerned in the transfer year under the Protocol minus the production allowances transferred,

(B) the maximum production level permitted for the substance or substances concerned in the transfer year under applicable domestic law minus the production allowances transferred, or

(C) the average of the actual national production level of the substance or substances concerned for the three years prior to the transfer minus the production allowances transferred.

(2) The United States may acquire production allowances from another Party if, at the time of such transfer, the Administrator finds that the other Party has revised its domestic production limits in the same manner as provided with respect to transfers by the United States in subsection (a).

(b) EFFECT OF TRANSFERS ON PRODUCTION LIMITS. - The Administrator is authorized to reduce the production limits established under this Act as required as a prerequisite to transfers under paragraph (1) of subsection (a) or to increase production limits established under this Act to reflect production allowances acquired under a transfer under paragraph (2) of subsection (a).

(c) REGULATIONS. - The Administrator shall promulgate, within two years after the date of enactment of the Clean Air Act Amendments of 1990, regulations to implement this section.

(d) DEFINITION. - In the case of the United States, the term 'applicable domestic law' means this Act.

SEC. 617. INTERNATIONAL COOPERATION

(a) IN GENERAL. - The President shall undertake to enter into international agreements to foster cooperative research which complements studies and research authorized by this title, and to develop standards and regulations which protect the stratosphere consistent with regulations applicable within the United States. For these purposes the President, through the Secretary of State and the Assistant Secretary of State for Oceans and International Environmental and Scientific Affairs, shall negotiate multilateral treaties, conventions, resolutions, or other agreements, and formulate, present, or support proposals at the United Nations and other appropriate international forums and shall report to the Congress periodically on efforts to arrive at such agreements.

(b) ASSISTANCE TO DEVELOPING COUNTRIES. - The Administrator, in consultation with the Secretary of State, shall support global participation in the Montreal Protocol by providing technical and financial assistance to developing countries that are Parties to the Montreal Protocol and operating under article 5 of the Protocol. There are authorized to be appropriated not more than $30,000,000 to carry out this section in fiscal years 1991, 1992 and 1993 and such sums as may be necessary in fiscal years 1994 and 1995. If China and India become Parties to the Montreal Protocol, there are authorized to be appropriated not more than an additional $30,000,000 to carry out this section in fiscal years 1991, 1992, and 1993.

SEC. 618. MISCELLANEOUS PROVISIONS

For purposes of section 116, requirements concerning the areas addressed by this title for the protection of the stratosphere against ozone layer depletion shall be treated as requirements for the control and abatement of air pollution. For purposes of section 118, the requirements of this title and corresponding State, interstate, and local requirements, administrative authority, and process, and sanctions respecting the protection of the stratospheric ozone layer shall be treated as requirements for the control and abatement of air pollution within the meaning of section 118.

Written by the Environmental Protection Agency's (EPA) Stratospheric Protection Division.

SECTION 4

608—TECHNICIAN CERTIFICATION PROGRAMS

This list of technician certification programs is up to date at the time of publication. It must be recognized, however, that it is subject to change from time to time. Some programs may be deleted and other programs may be added.

Programs that appear in this Appendix are approved by the Environmental Protection Agency (EPA) to provide the technical certification test. The EPA, however, does not review, approve, or recommend any particular training program or training material.

The training programs are listed by state. The addresses and phone numbers listed are for each program's headquarters. Many of the testing locations are located throughout the United States, not just in the headquarter's state. Each program should be able to provide you with a list of the testing schedules and test locations. Many even offer a test-by-mail for the small appliance certification.

The cost of the certification tests varies widely. The tests may be free for some who are enrolled in a related course in selected colleges and universities or may be discounted by as much as 45 percent in other cases. Some vocational–technical schools may include the cost of the test in the tuition for those taking related courses. Trade associations generally discount their fees to members by as much as 33 percent.

Programs charge a flat fee ranging from $20 to $150 regardless of how many test types are attempted. Some programs charge a registration fee plus a fee for each type attempted; others charge a flat fee for each type test attempted, ranging from $10 to $110.

Generally, retesting fees are the same as the initial testing fees. Some, however, reduce their retesting fee by as much as 40 percent.

A number of the programs offer mail-in testing for Type I Small Appliance Certification. These taking these open-book tests are required to achieve a higher percentage of correct responses to earn a certificate. Although 70 percent is generally the passing score for a proctored closed-book test, mail-in exams require a score of 80 percent, or better, to be successful.

The following programs are given alphabetically, by the state in which they are headquartered.

Programs

Alaska

Seward   (99664)
   Alaska Vocational Technical Center
   P.O. Box 889
   809 Second Avenue
   (907) 224-3322

Arizona

Phoenix   (85021)
   Motorcoach Training & Development
   9850 North 19th Drive, Suite 1
   (800) 255-2122

                 (85034)
   The Refrigeration School, Inc.
   4210 East Washington Street
   (602) 275-7133

                 (85017)
   Universal Technical Institute
   3823 North 34th Avenue
   (602) 271-4174

California

Freemont   (94539)
   Sequoia Institute
   200 Whitney Place
   (510) 490-6900

Lodi   (95241)
   Hartsog Trade School, Inc.
   P.O. Box 760
   (209) 339-9324

Los Angeles   (90057)
   Operating & Maintenance Engineer Trade
   Training Trust Fund for California & Nevada
   2501 West Third Street
   (213) 385-2889

Milpitas   (95035)
  Advanced Technical Training
  45 South Victoria, Suite 139
  (408) 534-3139

San Diego   (92101)
  San Diego City College
  1313 12th Avenue
  (619) 230-2080

San Jose   (95128)
  San Jose City College
  Applied Science Division
  2100 Moorpark Avenue, Room 304
  (408) 288-3781

Stockton   (95207)
  San Joaquin Delta College
  5151 Pacific Avenue
  (209) 474-5230

Colorado
Denver   (80210)
  Technology Training Inc.
  2111 South Adams Street
  (303) 759-2471

Delaware
Dover   (19903)
  Central Delaware Training Academy, Inc.
  P.O. Box 1344
  (302) 677-1534

Newark   (19713)
  Delaware Technical and Community College
  400 Stanton-Christian Road
  (362) 453-3001

Wilmington   (19801)
　Delaware Skills Center
　Building Maintenance
　13th & Poplar Streets
　(302) 654-5392

District of Columbia

Washington (DC)(20009)
　Air Conditioning Contractors of America
　Ferris State University
　1712 New Hampshire Avenue, NW
　(202) 483-9370

Ferris State University
See: Air Conditioning Contractors of America

(20001)
United Association of Journeymen and Apprentices of the Plumbing and Pipe Fitting Industry of the United States
901 Massachusetts Avenue, NW
(202) 628-5823

Florida

Boca Raton   (33478)
　Gables Residential
　6551 Park of Commerce Boulevard, Suite 100
　(561) 997-9700

Gainesville   (32653)
　National Assessment Institute/Block & Associates
　2100 Northwest 53 Avenue, Suite 1303
　(800) 280-3926

Orlando   (32819)
　Vidal Enterprises
　5001 Apopka Vineland Road
　(900) 258-4325

Punta Gorda   (33951)
  Thunder-Light, Inc.
  P.O. Box 1001
  (941) 637-8537

Rockledge   (32955)
  Mainstream Engineering Corporation
  Pines Industrial Center
  200 Yellow Place
  (407) 631-3550

Georgia

Atlanta   (30340)
  Association of Energy Engineers
  4025 Pleasantdale Road, Suite 420
  (404) 447-5083, ext. 215

Illinois

Buffalo Grove   (60089)
  Telemedia, Inc.
  750 Lake Cook Road
  (806) 837-3155

Champaign   (61821)
  Parkland College
  Business Training Center
  2400 West Bradley Avenue
  (217) 351-2281

Chicago   (60614)
  Coyne American Institute
  1235 West Fullerton Avenue
  (312) 935-2520

Des Plaines   (60016)
  Refrigeration Service Engineers Society
  1666 Rand Road
  (847) 297-6464

608–Technician Certification Programs

Joliet   (60432)
  Joliet Junior College
  IET/Workforce Development Center
  214 North Ottawa Street
  (815) 727-6544, ext. 1317

Lumbard   (60148)
  Association of Home Appliance Manufacturers
  North American Retail Dealers of America
  10 East 22nd Street, Suite 310
  (630) 953-8956

               (60148)
  Technical Seminars
  P.O. Box 995
  (800) 647-0385

Mt. Prospect   (60056)
  ESCO Institute
  1350 West Northwest Highway, Suite 205
  (800) 726-9696

Palatine   (60067)
  William Rainey Harper College
  CAD and Manufacturing Center
  1200 West Algonquin Road
  (847) 925-6000

Rockford   (61114)
  Rock Valley College
  3301 North Mulford Road
  (815) 654-4250

South Holland   (60473)
  Pace Maintenance and Technical Services Department
  Pace Acceptance Facility
  405 Taft Drive
  (708) 331-9127

Springfield (62794)
  Lincoln Land Community College
  Shepherd Road
  (217) 789-2200

Indiana
Indianapolis (46226)
  AC/C Tech
  4415 Forest Manor Avenue
  (317) 545-7071

Kansas
Wichita (67219)
  Climate Control Institute, Inc.
  3030 North Hillside
  (316) 686-7355

Maryland
Hanover (21076)
  Refrigeration Environmental Protection Association
  7525-M Connelley Drive
  (800) 435-3331

Linwood (21791)
  Environmental Training Group, Inc.
  428 McKinstry's Mill Road
  (800) 453-4200

Piney Point (20674)
  Seafarer's Harry Lundeberg School of Seamanship
  P.O. Box 75
  State Route 249
  (301) 994-0010, ext 274

Massachusetts
Canton (02021)
  Bay State School of Appliances
  225 Turnpike Street (Route 138)
  (617) 828-3434

Stoneham   (02180)
  Associated Technical Institute
  P.O. Box 31
  171 Main Street, Suite 30
  (718) 279-2280
  (800) 229-1284

Michigan

Battle Creek   (49017)
  Kellogg Community College
  Regional Manufacturing Technology Center
  450 North Avenue
  (616) 965-4137, ext. 2813

Sanford   (48657)
  Technical Training Services
  3233 Douglas Drive
  (517) 687-7637

Missouri

St. Louis   (63122)
  Metropolitan Manufacturers Association
  10733 Big Bend Boulevard
  (314) 966-1006

              (63313)
  Ranken Technical College
  4431 Finney Avenue
  (314) 371-0236

              (63127)
  Vatterott College
  Director: HVAC Programs
  12970 Maurer Industrial Drive
  (314) 843-4200

Fenton   (63026)
  Union Electric Company
  Power Plants Training Center
  1599 Fenpark Drive
  (314) 992-7422

Nebraska

Omaha   (68102)
  Universal Technical Institute
  902 Capitol Avenue
  (402) 345-2422

Nevada

Las Vegas   (89109)
  Air Conditioning Training by Quality
  3141 Westwood Drive
  (702) 731-1617

Henderson   (89015)
  Community College of Southern Nevada
  700 South College Drive, HIA
  (702) 564-7484, ext. 237

New Jersey

Blackwood   (08012)
  Pennco Tech
  P.O. Box 1427
  99 Erial Road
  (609) 232-0310

New York

Bronx   (10465)
  State University of New York
  Maritime College
  6 Pennyfield Avenue
  (718) 409-7340 or 409-2013

Brooklyn   (11211)
  H.V.A.C. Tech Inc.
  136 Metropolitan Avenue
  (718) 388-026

Lockport   (14094)
Niagara County Community College
Division of Lifelong Learning
Department of Corporate Training
136 Walnut Street
(716) 433-1836

New York City   (10001)
ACI—Environmental Safety Training Institute
239 West 29th Street, Ground Floor
(212) 254-5410

(10011)
Apex Technical School, Inc.
635 Avenue of the Americas
(212) 645-3300

(10001)
Refrigeration Training Center
208 West 29th Street, Suite 410
(212) 629-8520

(10001)
Technical Career Institutes
The College for Technology
320 West 31st Street
(212) 594-4000

New Mexico

Albuquerque   (87110)
ADC Limited
1919 San Mateo Northeast
(505) 265-5822

North Carolina

Raleigh   (27605)
North Carolina State Board of Refrigeration Examiners
P.O. 10666
(919) 755-5022

Ohio

Barberton   (44203)
Power Safety International
(*See* Technical Training Services)

Technical Training Services
20 South Van Buren Avenue
(800) 222-2625, ext. 1004

Niles   (44446)
ETI Environmental Control Program
2076 Youngton-Warren Road
(330) 652-9919

Toledo   (43606)
University of Toledo
Community and Technical College
(419) 531-3313

Oklahoma

Okmulgee   (74447)
Oklahoma State University/Okmulgee
1801 East Mission
(918) 756-6211, ext. 270

Pennsylvania

Apollo   (15690)
Center for Safety & Environmental Management
718 Jackson Road
(412) 478-2736

GQ Environmental, Inc.
(*See* Center for Safety & Environmental Management)

Media   (19063)
Delaware County Community College
901 South Media Line Road
(610) 359-5338

608–Technician Certification Programs

Scranton (18508)
Career Technology Center of Lackwanna County
Henry J. Dende Campus
3201 Rockwell Avenue
(717) 346-8471

Tennessee

Counce (38326)
Tennessee Valley Technical Programs
Route 1, Box 372
(901) 373-3992

Sparta (38583)
Educational Services
317 Fairview Circle
(931) 761-5024

Texas

Abilene (79604)
CFC Reclamation and Recycling Service, Inc.
P.O. Box 560
(915) 675-5311

Alice (78332)
National Certification Institute
425 South Reynolds
(512) 664-9602

Arlington (76011)
VGI Training Division
Video General Inc.
1156 107th Street
(917) 640-8333

Corpus Christi (18480)
Training Specialists
P.O. Box 181075
(512) 949-9780

San Antonio (78223)
Texas Engineering Extension Service
The Texas A&M University System
9350 South Presa
(210) 633-1000

Waco (76705)
Texas State Technical College at Waco
Air Conditioning and Refrigeration Technicians
3801 Campus Drive
(817) 867-4850

Utah

Provo (84603)
Geneva Steel
P.O. Box 2500
(801) 227-9000

Virginia

Arlington (26203)
Air-Conditioning & Refrigeration Institute
4301 North Fairfax Drive, Suite 425
(703) 524-8800

Norfolk (23510)
Marine Safety Consultants/
Tidewater School of Navigation, Inc.
100 W. Plume Street, Suite 450
(757) 625-7004

(23505)
Unified Industries Incorporated
7460 Tidewater Drive, Suite 100
(757) 480-1642

Falls Church (22041)
National Association of Power Engineers
Education Foundation
5707 Seminary Road, Suite 200
(703) 845-7055, ext. 12

(22046)
National Association of Plumbing-Heating-Cooling Contractors
P.O. Box 6808
180 South Washington Street
(800) 533-7694/(703) 237-8100

Washington

District of Columbia
(*See* District of Columbia, Washington)

Seattle   (98107)
Seattle Maritime Academy
4455 Shilshole Avenue, Northwest
(206) 782-2647

West Virginia

Morgantown   (26506)
West Virginia University Extension Service
Safety and Health Extension
130 Tower Lane
P.O. Box 6615
(304) 293-3096

Wisconsin

Waukesha   (53188)
State of Wisconsin
Department of Commerce
401 Pilot Court, Suite C
(414) 548-8617

Wyoming

Laramie   (82072)
F & J Air Conditioning & Refrigeration
Training and Certification
4421 Crow Drive, #B
(307) 742-4607

SECTION 5

GLOSSARY

Absolute humidity: The weight, in grains of water vapor, actually contained in 1 cubic foot (0.0283 cubic meter) of air.

Absolute pressure: The sum of gauge pressure plus atmospheric pressure.

Absolute temperature: The temperature at which all molecular motion of a substance stops and the substance theoretically contains no heat.

Absorbent: A substance that has the ability to absorb another substance.

Absorption refrigeration: A refrigeration system using lithium bromide or other refrigerant that loses and gains heat due to a change of state without the requirement for a large amount of compression.

Absorptivity: The ratio of radiant energy absorbed by an actual surface at a given temperature to that absorbed by a black body at the same temperature.

Acceptable: A Significant New Alternatives Policy (SNAP) designation meaning that a substitute refrigerant may be used, without restriction, to replace the relevant ozone-depleting substance (ODS) within the end-use specified. For example, R-22 is an acceptable substitute for R-502 in industrial process refrigeration. Note that all SNAP determinations apply to the use of a specific product as a substitute for a specific ODS in a specific end use.

Acceptable Subject to Narrowed Use Limits: A Significant New Alternatives Policy (SNAP) designation meaning that a substitute would be unacceptable unless its use was restricted to specific applications within an end use. This designation is generally used when the specific characteristics of different applications within an end use result in differences in risk. Use of the substitute in the end use is legal only in those applications included within the narrowed use limit. Note that all SNAP determinations apply to the use of a specific product as a substitute for a specific ozone-depleting substance (ODS) in a specific end use.

Acceptable Subject to Use Conditions: A Significant New Alternatives Policy (SNAP) designation meaning that a substitute would be unacceptable unless it is used under certain conditions. An example is the set of use conditions placed on motor vehicle air-conditioning refrigerants, requiring the use of unique fittings and labels and requiring that the original refrigerant be removed before charging with an alternative. Use of the substitute in the end use is legal provided the conditions are fully met.

Accessible hermetic: A single unit containing both motor and compressor and may be field serviceable.

Accumulator: A tanklike vessel in the suction line of an air conditioning or refrigeration system to prevent liquid refrigerant from entering the compressor.

Acid condition: A condition in which the refrigerant and/or lubricant in the system is contaminated with other fluids that are acidic in nature.

ACR tubing: Tubing, usually copper (Cu), used in air conditioning and refrigeration (ACR). The ends are sealed and the tubing is clean and dehydrated.

Actuator: The part of a regulating valve that changes fluid or thermal or electric energy into mechanical motion to open or close devices such as valves and dampers.

Add-On Heat Pump: A heat pump that has been installed in conjunction with an existing fossil fuel furnace.

Adiabatic compression: The compression of a gas without the addition or removal of heat.

Aeration: The combining of a substance with air.

Agitator: A device used to induce motion in a confined fluid.

Air: A mixture of oxygen (O) and nitrogen (N) and slight traces of other gases that may also contain moisture (humidity).

Air change: The number of times per hour that air in a room is changed either by mechanical means or by infiltration of outside air.

Air cleaner: A device designed to remove airborne impurities such as dust, fumes, and smoke.

Air coil: The coils on refrigeration and air-conditioning units through which air is blown.

Air conditioning: The simultaneous control of the temperature, humidity, air motion, and air distribution within an enclosure.

Air-conditioning unit: A device used for the treatment of air consisting of a means for ventilation, circulation, cleaning, and heat transfer to maintain temperature within a prescribed limit.

Air diffuser: An air distribution element or outlet designed to direct the flow of air in a desired pattern.

Airflow: The distribution or movement of air.

Air handler: A unit consisting of the fan or blower, heat transfer element, filter, and housing components of an air distribution system.

Air infiltration: The leakage of air into a structure through cracks and crevices, doors, windows, and other openings.

Air-sensing thermostat: A thermostat with a sensing bulb located in the airstream that operates as a result of changing air temperature.

Air, standard: *See* Standard air.

Algae: A low form of plant life found in water, especially troublesome in water cooling towers.

Algicide: A microbicide product used to combat excessive growth and damage due to algae, bacteria, or fungi in cooling systems.

Alkalinity: The amount of bicarbonates, carbonates, and hydroxides in water.

Altitude correction: A change in atmospheric pressure at higher altitudes has an effect on a vapor pressure bellows and adjustments are made in control settings to correct for the effect of altitude.

Ambient temperature: The temperature of the air that surrounds an object.

Glossary

Ambient wet bulb temperature: Air temperature measured on the windward side of a tower, free from the influence of the tower.

Apparatus dew point: The dew point of the air leaving the air conditioning coil.

Approach temperature: The difference between temperature of cold water leaving the tower and the ambient wet-bulb temperature.

ASTM standards: Standards that are set by the American Society for Testing and Materials.

Atmospheric pressure: The pressure exerted by the atmosphere in all directions, as indicated by a barometer. Atmospheric pressure at sea level is 14.696 psig (101.33 kPa).

Atomize: To change a liquid to small particles or fine spray.

Automatic control: The control of various functions of a piece of equipment without manual adjustment.

Automatic defrost: A method of removing ice and frost from the evaporator automatically.

Automatic expansion valve (AEV): A pressure-controlled metering device that reduces the high-pressure liquid refrigerant to a low-pressure liquid refrigerant.

Azeotrope: A substance having constant maximum and minimum boiling points.

Azeotropic mixture: Refrigerants that do not combine chemically but provide the desired refrigerant characteristics.

Back pressure: The pressure in the low side of a refrigerating system, also called *suction pressure* or *low-side pressure*.

Back seating: The fluid opening/closing such as a gauge opening to seat the joint where the valve stem goes through the valve body.

Baffle: A plate or vane used to direct or control movement of fluid or air within a confined area.

Ball check valve: A ball and seat valve assembly that permits flow of fluid in one direction only.

Barometer: An instrument for measuring atmospheric pressure.

Baudelot cooler: A heat exchanger in which water flows by gravity over the outside of tubes or plates.

Bearing: A low-friction device used for supporting and aligning moving parts.

Bellows: A corrugated cylindrical container that expands and contracts as the pressure inside changes.

Belt: A rubberlike continuous loop placed between two or more pulleys to transfer rotary motion.

Bending spring: A spring that is placed on the inside or outside of a tube to prevent collapse during the bending process.

Bimetal strip: Temperature regulating or indicating device with two dissimilar metals with unequal expansion rates as temperatures change.

Blast freezer: Low-temperature evaporator that uses a fan to force air rapidly over the evaporator surface.

Bleed: To release pressure slowly from a system or cylinder by opening a valve slightly.

Bleed valve: A valve with a small opening through which pressure is permitted to escape at a slow rate when the valve is closed.

Bleeding: Slowly reducing the pressure of liquid or gas from a system or cylinder by slightly opening a valve.

Boiling point: The temperature at which a liquid boils.

Glossary

Bourdon tube: A thin-walled flattened circular-shaped tube of elastic metal used in pressure gauges.

Boyle's law: The law of physics that deals with the volume of gases as its pressure is varied. If the temperature remains constant, the volume varies; if the pressure is increased, the volume is decreased; if the pressure is reduced, the volume is increased.

Braze: To join metals with a nonferrous filler using heat between the temperatures of 800°F (427°C) and the melting point of the base metal.

British thermal unit (Btu): The quantity of heat required to raise the temperature of 1 pound (0.4535 kg) of water (H_2O) 1°F (0.5556°C).

Bypass: A pipe or duct usually controlled by a valve or a damper for short circuiting the flow of a fluid.

Calibrate: To position the pointers, needles, or make other adjustments needed to determine accurate measurements.

Calorie: A large heat unit equal to the amount of heat required to raise 2.205 pounds (1 kg) of water (H_2O) 1°C (1.8°F).

Calorimeter: A device that is used to measure quantities of heat or determine specific heats.

Cam: An oblong mechanical component that gives a reciprocating motion when rotated.

Capacity: The heat absorbing ability, per unit of time, usually measured in British thermal units per hour.

Capillary tube: A small-diameter tube used to regulate the flow of liquid refrigerant.

Carbon filter: Air filter using activated carbon as an air cleansing agent.

Carrene: A refrigerant in Group 1 known as R-11. Chemical combination of carbon (C), chlorine (Cl), and fluorine (F).

Cascade systems: An arrangement whereby two or more refrigeration systems are used in series; the evaporator of one unit cools the condenser of another unit to produce ultra-low temperatures.

Cavitation: A gaseous condition found in low-pressure places of a liquid stream.

Celsius: A temperature scale in which the freezing point of water (H_2O) is 0° and the boiling point is 100°.

Centimeter: A metric unit of linear measurement equal to 0.3937 inch.

Centrifugal compressor: A pump that compresses gaseous refrigerants by centrifugal force.

Change of state: A condition in which a substance changes from a solid to a liquid, a liquid to a gas, a gas to a liquid, or a liquid to a solid, due to adding or removing heat.

Charge: The amount of refrigerant or lubricant in a system.

Charles's law: Volume of a given mass of gas at a constant pressure varies according to its temperature.

Check valve: A device that permits fluid flow in one direction only.

Chemical refrigeration: System of cooling using a disposable refrigerant. Also called an *expendable refrigerant system*.

Chill factor: A calculated number based on actual temperature and wind velocity.

Chiller: An air-conditioning system that circulates chilled water to various cooling coils in an installation.

Chimney effect: Tendency of air or gas to rise when heated.

Chlorofluorocarbon (CFC): Compound consisting of chlorine (Cl), fluorine (F), and carbon (C) atoms, which are very stable in the troposphere.

Glossary

Choke tube: A throttling device, such as a capillary tube, used to maintain correct pressure difference between high and low side in refrigerating mechanisms.

Circuit: Tubing, piping, or electrical wire installation that permits flow to and from the energy source.

Clearance pocket compressor: A small space in a cylinder from which compressed gas is not completely expelled.

Code installation: Refrigeration or air conditioning installation that conforms to the local code and/or the national code for safe and efficient installations.

Coefficient of conductivity: A measure of the relative rate at which different materials conduct heat.

Coefficient of expansion (COE): An increase in unit length, area, or volume for each degree rise in temperature.

Coefficient of Performance (COP): The ratio of work performed or accomplished as compared with the energy used.

Cogeneration: A primary source of energy that is also used to produce a secondary source of energy such as the use of waste heat from an electrical energy generation system to heat a building.

Cold: The absence of heat.

Cold junction: The part of a thermoelectric system that absorbs heat as the system operates.

Combustible liquids: A class 3 liquid having a flash point at or above 140°F (60°C).

Comfort chart: A chart used in air conditioning to show the dry bulb temperature, humidity, and air movement for human comfort conditions.

Comfort cooler: A system used to reduce the temperature in the living space but does not provide complete control of heating, humidifying, dehumidification, and air circulation.

Comfort cooling: Refrigeration for comfort as opposed to refrigeration for food storage.

Comfort zone: Area on psychometric chart that shows conditions of temperature, humidity, and sometimes air movement at which most people are comfortable.

Commercial refrigeration: A reach-in or service refrigerator of commercial size with or without the means of cooling.

Compound gauge: An instrument used to indicate the pressure both above and below atmospheric pressure. Commonly used to measure low-side pressures.

Compression: A term used to denote increase of pressure on a fluid by using mechanical energy.

Compression gauge: An instrument used to measure pressures above atmospheric.

Compression ratio: (1) Ratio of the volume of the clearance space to the total volume of the cylinder. (2) The ratio of the absolute low-side pressure to the absolute high-side pressure.

Compression system: A refrigeration system in which the pressure-imposing element is mechanically operated, as opposed to an absorption system that has no compressor.

Compressor: The pump of a refrigerating mechanism that draws a low pressure on cooling side of refrigerant cycle and squeezes or compresses the gas into the high-pressure or condensing side of the cycle.

Compressor displacement: The volume in inches or millimeters of the area of the piston head multiplied by the length of the piston stroke and the number of pistons.

Compressor, external drive: *See* Compressor, open type

Compressor, hermetic: Compressor in which the driving motor is seated in the same dome or housing as the compressor.

Glossary

Compressor, multiple stage: A compressor having two or more compressive steps. Discharge from each step is the intake pressure of the next in the series.

Compressor, open type: A compressor in which the crankshaft extends through the crankcase and is driven by an outside motor.

Compressor, reciprocating: A compressor that uses a piston and cylinder mechanism to provide pumping action.

Compressor, rotary: A compressor that uses vanes, eccentric mechanisms, or other rotating devices to provide pumping action.

Compressor seal: A leak-proof seal between crankshaft and compressor body in open-type compressors.

Compressor, single-stage: A compressor having only one compressive step between low-side pressure and high-side pressure.

Condensate: Moisture from the air as a result of the removal of heat from a vapor to bring it below the dew-point temperature.

Condensate pump: A device used to remove water condensate that collects beneath an evaporator coil.

Condensation: Liquid or droplets that form when a gas or vapor is cooled below its dew point.

Condense: Action of changing a gas or vapor to a liquid.

Condenser, air-cooled: A heat exchanger that transfers heat to the surrounding air.

Condenser comb: A comblike device, metal or plastic, used to straighten the metal fins on condensers or evaporators.

Condenser fan: The fan that forces air over the condenser coil.

Condenser, water-cooled: A heat exchanger designed to transfer heat from hot gaseous refrigerant to water.

Condensing pressure: The pressure inside the condenser at which the refrigerant vapor gives off latent heat of vaporization and changes to a liquid.

Condensing temperature: The temperature inside the condenser at which the vaporous refrigerant gives off latent heat of vaporization and becomes a liquid.

Condensing unit: The part of a refrigeration mechanism that pumps vaporized refrigerant from the evaporator, compresses it, liquefies it in the condenser, and returns it to the refrigerant metering device.

Condensing unit service valve: A shut-off valve found on the condensing unit to enable the technicians to service the unit.

Conduction: The flow of heat between substances by molecular vibration.

Conductivity: The amount of heat in British thermal units (calories), transmitted in 1 hour through 1 foot (0.3 m) of a homogeneous material 1 inch (25.4 mm) thick for a difference in temperature of 1°F (0.556°C) between the surfaces of the material.

Constrictor: A tube or orifice used to restrict flow of a gas or a liquid.

Contaminant: A substance such as dirt, moisture, or some other material that is foreign to the refrigerant or the oil in the system.

Continuous cycle absorption system: A system that has a continuous flow of energy input.

Control: Any device used for the regulation of a machine in normal operation, manual or automatic, usually responsive to temperature or pressure.

Control, defrosting: A device used to automatically defrost an evaporator.

Controller: A group of controls and circuits used to accurately and automatically operate a system or device.

Control, low-pressure: A cycling device connected to the low-pressure side of system.

Glossary

Control, motor: A temperature or pressure-operated device used to control running of motor.

Control, refrigerant: A device used to regulate flow of liquid refrigerant into evaporator.

Control system: All of the components required for the automatic control of a process variable.

Control, temperature: A temperature-operated thermostatic device that automatically opens or closes a circuit.

Control valve: A valve that regulates the flow or pressure of a medium that affects a controlled process.

Convection: The transmission of heat by the circulation of a liquid or a gas such as air.

Convection, forced: The transfer of heat resulting from forced movement of liquid or gas by means of a fan or pump.

Convection, natural: The circulation of a gas or liquid due to the difference in density resulting from temperature differences.

Convector: A surface designed to transfer its heat to the surrounding air by means of convection.

Cooler: A heat exchanger that transfers heat from one substance to another.

Cooling tower: A device that cools by water evaporation in air. The water is cooled to the wet-bulb temperature of the air.

Copper plating: An abnormal condition that exists when moisture is present in a refrigeration system, resulting of copper being electrolytically deposited on the steel parts of the system.

Corrosion: Deterioration of materials from chemical action.

Counterflow: The opposing direction of flow of fluids, the coldest portion of one meeting with the warmest portion of the other.

Coupling: A mechanical device used to join refrigerant lines.

"Cracking" a valve: Opening a valve a small amount.

Crankshaft seal: Leak-proof joint between the crankshaft and the compressor body.

Crank throw: The distance between the center line of a main bearing journal and the center of the crankpin or eccentric.

Critical pressure: The compressed condition of a refrigerant at which the vapor and liquid have the same properties.

Critical temperature: The temperature at which vapor and liquid have the same properties.

Critical vibration: A vibration that is noticeable and harmful to structure.

Cross charged: A sealed container of two fluids that together create a desired pressure-temperature curve.

Cryogenic fluid: A substance that exists as a liquid or gas at ultra low temperatures –250°F (–157°C) or lower.

Cryogenics: Refrigeration that deals with producing below-zero temperatures, –250°F (–157°C) and lower.

Cut-in: The temperature or pressure value at which the control circuit closes.

Cut-out: The temperature or pressure value at which the control circuit opens.

Cycle: The complete course of operations of a refrigeration system, including the four major functions: compression, condensation, expansion, and evaporation.

Cylinder: (1) The chamber in a compressor in which the piston travels to compress the refrigerant vapor. (2) Device that converts fluid power into linear mechanical force and motion. (3) A closed container for fluids.

Glossary

Cylinder head: A cap or plate that encloses the open end of a cylinder.

Cylinder, refrigerant: A tanklike container in which refrigerant is stored and dispensed.

Dalton's law: A law that states that a vapor pressure exerted in a container by a mixture of gases is equal to the sum of the individual vapor pressures of the gases contained in the mixture.

Damper: A device used to control the flow of air.

Deaeration: The act of separating air from a substance.

Decibel: A unit commonly used to express sound or noise intensity.

Decomposition: Spoilage.

Defrost: To remove accumulated frost and ice.

Defrost cycle: The process of removing ice or frost buildup from the coil.

Defrost timer: A device connected into the electric circuit to start the defrost cycle and keep it on until the ice has melted.

Degree: A unit of measure on a temperature scale.

Degree day: A unit that represents one degree of difference between the inside temperature and the average outdoor temperature for one day.

Degree of superheat: The difference between the boiling point of the refrigerant and the actual temperature above the boiling point.

Dehumidification: The reduction or removal of water vapor in the air.

Dehumidifier: A device used to lower the moisture content of the air passing through it.

Dehumidify: To remove water or moisture from the atmosphere; to remove water vapor or moisture from any material.

Dehydrated oil: A lubricant from which the moisture has been removed to an acceptable level.

Deice control: A device used to control the compressor and allow for the melting of any accumulation of frost on the evaporator.

Density: The weight per unit volume of a substance.

Desiccant: A substance used to collect and hold moisture in a refrigeration system.

Design pressure: The highest pressure expected to be reached during normal operation, usually the operating pressure plus a safety factor.

Dew point: The temperature at which a vapor begins to condense, usually at 100 percent relative humidity.

Diaphragm: A flexible material, usually a thin metal or rubber, used to separate chambers.

Dichlorodifluoromethane: A CFC refrigerant commonly known as R-12.

Differential: The difference between the cut-in and cut-out of a pressure or temperature control.

Direct expansion evaporator: An evaporator that uses either an automatic or a thermostatic expansion valve.

Draft gauge: An instrument used to measure air movement by measuring the air pressure differences.

Drier: A device containing a desiccant used to remove moisture from refrigerant.

Drip pan: A pan or trough used to collect condensate from an evaporator coil.

Dry bulb: A thermometer used to measure the ambient air temperature.

Dry-bulb temperature: The actual temperature of the air, as opposed to wet-bulb temperature.

Dry ice: Compressed carbon dioxide.

Dry system: A refrigeration system in which only droplets are present in the evaporator.

Duct: A pipe or closed conduit of sheet metal fiberglass board or other suitable material used in conducting air to and from an air handling unit.

Eccentric: A disk, wheel, or cam having an axis that is displaced from its center so that it is capable of imparting a reciprocating motion when turned.

Effective temperature: The overall effect of the air temperature, humidity, and air movement on a human.

Electronic leak detector: An electronic instrument that senses refrigerant vapor in the atmosphere.

End use: A Significant New Alternatives Policy (SNAP) definition for a process or class of specific applications within major industrial sectors where a substitute is used to replace an ozone-depleting substance. The specific definition varies by sector, but examples are motor vehicle air conditioning, electronics cleaning, flooding fire extinguishing systems, and polyurethane integral skin foam. Substitutes are listed by end use in the SNAP lists. In order of increasing specificity, a particular system is part of an industrial use sector, an end use, and an application.

Energy: The ability to do work.

Enthalpy: The actual or total heat contained in a substance.

Epoxy: An adhesive formed by mixing two resins.

Equipment room: *See* Machine room.

Eutectic: A mixture of two substances that provides the lowest melting temperature of all the various mixes of the two substances.

Evacuation: The removal of air and moisture from a system.

Evaporation: The change of state of a liquid to a gas.

Evaporative condenser: A specially designed condenser used to remove heat from refrigerant gas by using the cooling effect of evaporating water.

Evaporator: The part of a refrigeration system where the refrigerant boils and picks up the heat.

Evaporator fan: The fan that forces the air through the evaporator.

Environment: The conditions of the surroundings.

Exhaust valve: The outlet port that allows the compressed vapor to escape from the cylinder. Also called the *discharge valve.*

Expansion valve: A metering device in a refrigeration system that reduces a high-pressure liquid refrigerant to a low-pressure liquid refrigerant.

External equalizer: The tube connected to the low-pressure side of the diaphragm in the thermostatic expansion valve.

Fahrenheit scale: The scale on a standard thermometer, having the boiling point of water at 212°F and the freezing point at 32°F.

Filter: A device for removing dust particles from air, or unwanted elements from liquids.

Fine: A gas or air passage through which the products of combustion escape to the atmosphere.

Finned tube: A tube built up with an extended surface in the form of fins.

Fitt: The sheet metal extension on evaporator or condenser tubes.

Flammability: The ability to burn.

Glossary

Flammable liquids: Liquids whose flash points are below 140°F (60°C) and whose vapor pressures are less than 40 psig (276 kPa) at 100°F (37.8°C).

Flapper valve: A thin metal valve used as the suction and discharge valves in refrigeration compressors.

Flare: An enlargement, usually at a 45° angle, on the end of a piece of tubing used to connect it to a fitting or another piece of tubing.

Flare fitting: A type of soft tubing connector that requires the tube be flared to make a mechanical seal.

Flare nut: The fitting placed over the tubing to clamp the flared tubing against another fitting.

Flash gas: The gas that is the result of the instantaneous evaporation of refrigerant in a metering device.

Gas: The vapor state of a substance.

Gasket: A resilient or flexible material used between mating surfaces to provide a leak-proof seat.

Gauge: An instrument used for measuring pressures both above and below atmospheric pressure.

Gauge manifold: A manifold that holds both the pressure and compound gauges, the valves that control the flow of fluids through the manifold ports, and the charging hose connections.

Gauge port: The opening or connection provided for the installation of gauges.

Gauge pressure: A measure of pressure taken with a gauge, measured from atmospheric as opposed to absolute.

Ground coil: A heat exchanger that is buried in the ground; usually the outdoor coil on a heat pump system.

Global warming: An increase in the natural greenhouse effect, refers to the physical phenomenon that may lead to heating of Earth.

Global warming potential: The relative ability of each greenhouse gas emission to affect radiative forcing and thereby the global climate.

gph or GPH: An abbreviation for gallons per hour.

Greenhouse effect: The light and heat from the sun that cannot be reflected through the atmosphere.

Greenhouse gases: Gases that are present in relatively small quantities in the atmosphere and strongly absorb infrared radiation or "heat" emitted by Earth.

GWP: An abbreviation for global warming potential.

Halide refrigerant: The family of synthetic refrigerants that contain halogen chemicals.

Halide torch: A device used to detect certain refrigerant leaks in a system.

Halogens: Substances that contain fluorine, chlorine, bromine, and iodine.

Head: A unit of pressure usually expressed in feet of water.

Header: A length of pipe large enough to carry the total volume to which several pipes are connected, used to carry fluid to the various points.

Head pressure: The pressure against which the compressor must deliver the gas.

Head pressure control: A pressure-operated control that opens the electric circuit when the head pressure exceeds preset limits.

Head velocity: The height of a flowing fluid that is equivalent to its velocity pressure.

Heat: A form of energy produced through the expenditure of another form of energy.

Heat content: The amount of heat, usually stated in British thermal units (calories) per pound, absorbed by a refrigerant in raising its temperature from a predetermined level to a final condition and temperature.

Heat exchanger: Any device that removes heat from one substance and adds it to another.

Heat gain: The amount of heat gained, measured in British thermal units, from a space to be conditioned, at the local summer outdoor design temperature and a specified indoor design condition.

Heating coil: A coil of piping used to transfer heat from a liquid.

Heating control: Any device that controls the transfer or heat.

Heating Seasonal Performance Factor (HSPF): The total heating output of a heat pump during its normal annual usage period for heating divided by the total electric power input in watt-hours during the same period.

Heating value: The amount of heat that may be released by the expenditure of energy.

Heating, ventilating, and air conditioning (HVAC): A term often used to describe the industry that produces the equipment that provides climatic comfort.

Heat intensity: The heat concentration in a substance.

Heat lag: The amount of time required for heat to travel through a substance heated on one side and not on the other.

Heat leakage: The flow of heat through a substance when a difference in temperature exists.

Heat load: The amount of heat, measured in British thermal units (calories) or watts, that must be added or removed by a piece of equipment during a 24-hour period.

Heat loss: A decrease in the amount of heat contained in a space, resulting from heat flow through walls, windows, roof, and other building surfaces and from exfiltration of warm air.

Heat of compression: The heat developed within a compressor when a gas is compressed as in a refrigeration system.

Heat of fusion: The amount of heat required to change a solid to a liquid or a liquid to a solid with no change in temperature.

Heat of the liquid: The heat content of a liquid or the heat necessary to raise the temperature of liquid from a predetermined level to a final temperature.

Heat of the vapor: The heat content of a gas or the heat necessary to raise the temperature of a liquid from a predetermined level to the boiling point plus the latent heat of vaporization necessary to convert a liquid to a gas.

Heat pump: A unit that both cools and heats.

Heat sink: A relatively cold surface that is capable of absorbing heat usually used as control points.

Heat transfer: The movement of heat from one body to another. The heat may be transferred by radiation, conduction or convection.

Heat unit: Usually refers to a British thermal unit (calorie).

Hermetic compressor: A unit in which the compressor and motor are sealed inside a housing.

High side: The part of the refrigeration system that contains the high-pressure refrigerant.

High-side charging: The process of introducing liquid refrigerant into the high side of a refrigeration system.

High-pressure cut-out: An electrical control switch operated by the pressure in the high-pressure side of the system that automatically opens an electric circuit when a predetermined pressure is reached.

High-vacuum pump: A vacuum pump capable of creating a vacuum in the range of 1000 to 1 microns.

Holding charge: A partial charge of refrigerant placed in a piece of refrigeration equipment after dehydration and evacuation either for shipping or testing purposes.

Horsepower: A unit of power. The effort necessary to raise 33,000 pounds (14,969 kg), a distance of one foot (0.304 m) in one minute.

Hot gas: The refrigerant gas leaving the compressor.

Hot-gas bypass: A connection from the compressor discharge directly into the suction side of a compressor.

Hot-gas defrost: A method of evaporator defrosting that uses the hot discharge gas to remove frost from the evaporator.

Hot-gas line: The line that carries the hot compressed vapor from the compressor to the condenser.

Hot junction: The part of a thermocouple that releases heat.

Hot water heating system: A heating system in which water is the medium to transport heat through pipes from the boiler to the heating units.

Humidifier: A device that adds moisture to warm air being circulated or directed into a space.

Humidistat: An automatic control that is sensitive to humidity and is used for the automatic control of relative humidity by reacting to changes in the moisture content of the air.

HVAC: An abbreviation for heating, ventilating, and air conditioning.

ICC: The abbreviation for the Interstate Commerce Commission.

Ice cream cabinet: A commercial refrigerator that maintains about 0°F (–17.8°C) temperature used for ice cream storage.

Idler: A pulley used on some belt drives to allow belt adjustment and to eliminate belt vibration.

Impeller: The rotating part of a pump that causes the water to flow.

Induced-draft cooling tower: A cooling tower in which the flow of air is created by one or more fans drawing the saturated air out of the tower.

Infiltration: The leakage of air into a building or space.

Inhibitor: A substance that prevents a chemical reaction such as corrosion or oxidation to metals.

Insulation, thermal: A material that has a high resistance to heat flow.

Interlock: A device that prevents certain parts of an air-conditioning or refrigeration system from operating when other parts of that system are not operating.

Interstate Commerce Commission (ICC): The government body that controls the design, construction, and shipping of pressurized containers.

Isothermal: A term describing a change in volume or pressure under constant-temperature condition.

Isothermal expansion and contraction: The expansion and contraction that takes place without a change in temperature.

Joint: The connecting point between two surfaces, as between two pipes.

Journal, crankshaft: The part of a crankshaft on which the bearing is in contact with the shaft.

Kelvin scale: A thermometer scale on which absolute zero is 0°.

King-valve: A service valve on the liquid receiver outlet.

Lag: A delay in the response to some demand.

Glossary

Lamp, sterile: A lamp with a high-intensity ultraviolet ray used in food storage cabinets and air ducts to kill bacteria.

Lantern gland (packing): A packing ring inside a stuffing box that has perforations for the introduction or removal of oil.

Lap: To smooth a metal surface to a high polish or accuracy using a fine abrasive.

Latent heat: The heat added to or removed from a substance to change its state but which cannot be measured by a change in temperature.

Latent heat of condensation: The heat removed from a vapor to change it to a liquid with no change in temperature.

Latent heat of vaporization: The quantity of heat required to change a liquid to a gas with no change in temperature.

Leak detector: A device or substance used to detect leaks, such as a halide torch, soap bubbles, dye, or by electronic means.

Liquid: A substance in which the molecules move freely among themselves, but do not tend to separate as in a vapor.

Liquid absorbent: A liquid chemical that has the ability to take on or absorb other fluids.

Liquid charge: Usually refers to the power element of temperature controls and thermostatic expansion valves.

Liquid filter: A very fine strainer used to remove foreign matter from the refrigerant.

Liquid indicator: A device located in the liquid line with a glass window through which the flow of liquid may be observed.

Liquid line: The line carrying the liquid refrigerant from the receiver or condenser-receiver to the evaporator.

Liquid receiver: A cylinder connected to the condenser outlet used to store liquid refrigerant.

Liquid receiver service valve: A two- or three-way manually operated valve located at the receiver outlet used for installation and service operations. Also called a *king valve*.

Liquid sight glass: A glass bull's-eye installed in the liquid line to permit visual inspection of the liquid refrigerant.

Liquid stop valve: A magnetically operated valve general used to control the flow of liquid to an evaporator; may also be used wherever on-off control is permissible.

Liquid strainer: *See* Liquid filter.

Liquid-vapor valve: A dual hand valve used on refrigerant cylinders to release either gas or liquid from the cylinder.

Liter: A metric unit of volume measurement equal to 0.2642 gallon.

Load: The required rate of heat removal. Heat per unit of time that is imposed on the system by a particular job.

Low-pressure control: A pressure-operated switch in the suction side of a refrigeration system that opens its contacts to stop the compressor at a given cut-out setting.

Low side: Composed of the parts of a refrigeration system in which the refrigerant pressure corresponds to the evaporator pressure.

Low-side charging: The process of introducing refrigerants into the low side of the system.

Low-side pressure: The pressure in the low side of the system.

Machine: A piece of equipment, sometimes the total unit.

Machine room: An area in which the refrigeration equipment, with the exception of the evaporator, is installed. Also referred to as *equipment room*.

Major maintenance: An Environmental Protection Agency (EPA) definition for maintenance, service, or repair that involves removal of the appliance compressor, condenser, evaporator, or auxiliary heat exchanger coil.

Glossary

Manifold: (1) The portion of the refrigerant main in which several branch lines are joined together. (2) A single piece in which there are several fluid paths. (3) The casting that is used to service a refrigerant system.

Manifold, discharge: A device used to collect compressed refrigerant from the various cylinders of a compressor.

Manifold, service: A chamber equipped with gauges, manual valves, and charging hoses used in servicing refrigeration units.

Manometer: An instrument for measuring small pressure. Also, a U-shaped tube partly filled with liquid.

Manual shut-off valve: A hand-operated device that stops the flow of fluids in a piping system.

Manual starter: A hand-operated motor switch equipped with an overload trip mechanism.

Master switch: The main switch that controls the starting and stopping of the entire system.

Mean effective pressure (MEP): The average pressure on a surface exposed to a varying pressure.

Mechanical efficiency: The ratio of work done by a machine to the energy used to do it.

Mechanical refrigeration: A term usually used to distinguish a compression system from an absorption system.

Melt: To change state from a solid to a liquid.

Melting point: The temperature, measured at atmospheric pressure, at which a substance will melt.

Mercury bulb: A glass tube that uses a small amount of mercury to make or break an electric circuit.

Met: A measure of the heat released from a human at rest.

Meter: (1) An instrument used for measuring. (2) A unit of length in the metric system.

Metric system: The decimal system of measurement.

Micrometer: A precision measuring instrument, with an accuracy of 0.001 to 0.0001 inch (0.0254 to 0.00254 mm).

Micron gauge: An accurate instrument used for measuring vacuum that is very close to a perfect vacuum.

Miscibility: The ability of several substances to be mixed together.

Modulating: Refers to a device that tends to adjust by small increments rather than being full-on or full-off.

Modulating control: A type of control system characterized by a control valve or motor that regulates the flow of air, steam, or water in response to a change in conditions at the controller. It operates on partial degree variations in the medium to which the controller is exposed.

Mollier's diagram: A graph indicating refrigerant pressure, heat, and temperature properties.

Monochlorodifluoromethane: A refrigerant that is better known as Freon-22 or R-22. Its chemical formula is $CHClF_2$; the cylinder color code is green.

Muffler: A device used in the hot-gas line to silence the compressor discharge surges.

Multiple system: A refrigeration system that has several evaporators connected to one condensing unit.

MVAC: Abbreviation for motor vehicle air conditioning.

MVAC-like appliance: An Environmental Protection Agency (EPA) definition for a mechanical vapor compression, open-drive compressor appliances used to cool the driver's or passenger's compartment of a nonroad vehicle, including agricultural and construction vehicles. This definition excludes appliances using HCFC-22.

Natural convection: The movement of a fluid caused by temperature differences.

Natural-draft cooling tower: A cooling tower in which the flow of air depends on natural air currents or a breeze.

Neoprene: A synthetic rubber that is resistant to refrigerants and oils.

Neutralizer: A substance used to counteract the action of acids.

Neutron: The core of an atom; it has no electrical potential (electrically neutral).

No-frost refrigerator: A low-temperature refrigerator cabinet in which no frost or ice collects on the evaporator surfaces or on the materials stored in the cabinet.

Nominal size tubing: Tubing whose inside diameter is the same as iron pipe of the same size.

Noncondensable gas: Any gas, usually in a refrigeration system, that cannot be condensed at the temperature and pressure at which the refrigerant will condense; it requires a higher head pressure.

Nonferrous: Metals and metal alloys that contain no iron.

Nonfrosting evaporator: An evaporator that never collects frost or ice on its surface.

Normal charge: A charge that is part liquid and part gas under all operating conditions.

Normally closed contacts: A contact pair that is closed when the device is in the energized condition.

Normally open contacts: A contact pair that is open when the device is in the energized condition.

O: Chemical symbol for oxygen.

O₃: Chemical symbol for ozone.

Odor: The contaminants in the air that affect the sense of smell.

Off cycle: The period when equipment, specifically a refrigeration system, is not in operation.

Oil binding: A condition in which a layer of oil on top of liquid refrigerant may prevent it from evaporating at its normal pressure and temperature.

Oil check valve: A check valve installed between the manifold and the crankcase of a compressor to permit oil to return to the crankcase but to prevent the exit of the oil from the crankcase when starting.

Oil, compressor lubricating: A highly refined lubricant made especially for refrigeration compressors.

Oil, entrained: Oil droplets carried by high-velocity refrigerant gas.

Oil equalizer: A pipe connection between two or more pieces of equipment made in such a way that the pressure, or fluid level, in each piece is maintained equally.

Oil filter: A device in the compressor used to remove foreign material from the crankcase oil before it reaches the bearing surfaces.

Oil level: The level in a compressor crankcase at which oil must be carried for proper lubrication.

Oil loop: A loop placed at the bottom of a riser to force oil to travel up the riser.

Oil pressure failure control: A device that acts to shut off a compressor whenever the oil pressure falls below a predetermined point.

Oil pressure gauge: A device used to show the pressure of oil developed by the pump within a refrigeration compressor.

Oil pump: A device that provides the source of power for force-feed lubrication systems in refrigeration compressors.

Oil return line: The line that carries the oil collected by the oil separator back to the compressor crankcase.

Oil separator: A device that separates oil from the refrigerant and returns it to the compressor crankcase.

Oil sight glass: A glass "bull's-eye" in the compressor crankcase that permits visual inspection of the compressor oil level.

Oil sludge: Usually a thick, slushy substance formed by contaminated oils.

Oil trap: (1) A low spot, sag in the refrigerant lines, or space in which oil will collect. (2) A mechanical device for removing entrained oil.

On cycle: The period when the equipment, specifically refrigeration equipment, is in operation.

Open circuit: An electric circuit that has been interrupted to stop the flow of electricity.

Open compressor: A compressor that uses an external drive.

Open display case: A commercial refrigerator cabinet designed to maintain its contents at the desired temperatures even though the cabinet is not closed.

Opening: An Environmental Protection Agency (EPA) definition for any service, maintenance, or repair on an appliance that would release class I or class II refrigerant from the appliance to the atmosphere unless the refrigerant were recovered previously from the appliance. Connecting and disconnecting hoses and gauges to and from the appliance to measure pressures within the appliance and to add refrigerant to or recover refrigerant from the appliance shall not be considered "opening."

Open-type system: A refrigeration system that uses a belt-driven or direct-coupling-driven compressor.

Operating cycle: A sequence of operations under automatic control intended to maintain the desired conditions at all times.

Operating pressure: The actual pressure at which the system normally operates.

Orifice: A precision opening used to control fluid flow.

Outdoor coil: Refrigerant-containing portion of a fan coil unit similar to a car radiator, typically made of several rows of copper tubing with aluminum fins.

Output: The amount of energy that a machine is able to produce in a given period of time.

Outside air, fresh air: Air from outside of the conditioned space.

Overload: A load greater than that for which the system or machine was designed.

Overload protector: A device designed to stop the motor if a dangerous overload condition occurs.

Overload relay: A thermal device that opens its contacts when the current through a heater coil exceeds the specified value for a specified time.

Oxidize: To burn, corrode, or rust.

Oxygen: An element in the air that is essential to animal life. The chemical symbol is O.

Ozone (O_3): An unstable pale blue gas. An allotropic form of oxygen (O) that is usually formed by a silent electrical discharge in the air.

Package unit: A complete set of refrigeration or air-conditioning components located in the refrigerated space.

Packing: (1) A resilient impervious material placed around the stems of certain types of valves to prevent leakage. (2) The slats or surfaces in cooling towers designed to increase the water to air contact.

Glossary

Packless valve: A valve that does not use a packing to seal around the valve stem.

Partial pressure: A condition occurring when two or more gases occupy a given space and each gas creates a part of the total pressure.

Parts per million (ppm): A unit of concentration in solutions.

Pascal's law: A law that states that a pressure imposed upon a fluid is transmitted equally in all directions.

Performance: A term frequently used to mean output or capacity, as performance data.

Performance factor: The ratio of heat removed by a refrigeration system to the heat equivalent of the energy required to do the job.

Pilot control: A valve arrangement used in an evaporator pressure regulator to sense the pressure in the suction line and to regulate the action of the main valve.

Pilot control, external: A method by which the internal connection of the pilot is plugged and an external connection is provided to make it possible to use an evaporator pressure regulator as a suction stop valve as well.

Pilot tube: A device that measures air velocities.

Piston: A disk that slides in a cylinder and is connected with a rod to exert pressure upon a fluid inside the cylinder.

Piston displacement: The volume displaced in a cylinder by the piston as it travels the full length of its stroke.

Pitch: The slope of a pipe line used to enhance drainage.

Plenum chamber: An air compartment to which one or more pressurized distributing ducts are connected.

Polystyrene: A plastic used as insulation in some refrigeration cabinets.

Power: The time rate of doing work.

Power element: The sensitive element of a temperature-operated control.

ppm: Abbreviation for parts per million.

Precooler: A cooler used to remove sensible heat before shipping, storing, or processing.

Pressure: The force exerted per unit of area.

Pressure, absolute: The pressure measured above an absolute vacuum.

Pressure, atmospheric: The pressure exerted by Earth's atmosphere.

Pressure, condensing: *See* Pressure, discharge.

Pressure, crankcase: The pressure in the crankcase of a reciprocating compressor.

Pressure, discharge: The pressure against which the compressor must deliver the refrigerant vapor.

Pressure drop: The loss of pressure due to friction of lift.

Pressure gauge: An instrument that measures the pressure exerted by the contents of a container.

Pressure, gauge: The pressure existing above atmospheric pressure.

Pressure, heat diagram: A graph representing a refrigerant's pressure, heat, and temperature properties. *Also* a Mollier's diagram.

Pressure, limiter: A device that remains closed until a predetermined pressure is reached and then opens to release fluid to another part of the system or opens an electric circuit.

Pressure motor control: A control that opens and closes an electric circuit as the system pressures change.

Glossary

Pressure-operated altitude valve (POA): A device used to maintain a constant low-side pressure regardless of the altitude of operation.

Pressure-regulating valve: A valve that maintains a constant pressure on its outlet side regardless of how much the pressure varies on the supply side of the valve.

Pressure regulator, evaporator: An automatic pressure-regulating valve installed in the suction line to maintain a predetermined pressure and temperature in the evaporator.

Pressure, saturation: The pressure at which a specific temperature is saturated.

Pressure, suction: The pressure forcing the gas to enter the suction inlet of a compressor.

Pressure tube: A small line carrying pressure to the sensitive element of a pressure controller.

Pressure water valve: A device that controls water flow through a water-cooled condenser in response to the heat pressure.

Primary control: A device that directly controls the operation of a fuel oil burner.

Quick-connect coupling: A device that permits fast and easy connection of two refrigerant lines by use of compression fittings.

Radiation: The transmission of heat through space by wave motion.

Ram air: The air forced through the radiator and condenser by the movement of an automobile along the road.

Range: A change within limits of the settings of pressure or temperature control.

Rankine scale: The name given to the absolute Fahrenheit scale. The zero point on this scale is $-459.67°F$.

Rating: The assignment of capacity.

Receiver, auxiliary: An extra vessel used to supplement the capacity of the receiver when additional storage volume is necessary.

Reciprocating: Term describing back-and-forth motion in a straight line.

Recirculated air: Air that is passed through the conditioning unit before being returned to the conditioned space.

Reclaim: An Environmental Protection Agency (EPA) definition to re-process refrigerant to at least the purity specified in the ARI Standard 700-1993, Specifications for Fluorocarbon Refrigerants, and to verify this purity using the analytical methodology prescribed in the Standard.

Recording thermometer: A temperature-sensing instrument that uses a pen to record on a piece of moving paper the temperature of a refrigerated space.

Recover: An Environmental Protection Agency (EPA) definition to remove refrigerant in any condition from an appliance and store it in an external container without necessarily testing or processing it in any way.

Recycle: An Environmental Protection Agency (EPA) definition to extract refrigerant from an appliance and clean refrigerant for reuse without meeting all of the requirements for reclamation. In general, recycled refrigerant is refrigerant that is cleaned using oil separation and single or multiple passes through devices, such as replaceable core filter-driers, which reduce moisture, acidity, and particulate matter.

Reed valve: A piece of thin, flat, tempered steel plate fastened to the valve plate.

Refrigerant circuit: Defined by the Environmental Protection Agency (EPA) for the parts of an appliance that are normally connected to each other (or are separated only by internal valves) and are designed to contain refrigerant.

Refrigerant: Substance used in refrigeration cycle that has a change of state from a liquid to a gas, and releases its heat in a condenser as the substance returns from the gaseous state to a liquid state.

Glossary

Refrigerant control: A control that regulates the flow of gas refrigerant from the high side to the low side of the refrigeration system.

Refrigerant tables: Tables that show the properties of saturated refrigerants at various temperatures.

Refrigerant velocity: The movement of gaseous refrigerant required to entrain oil mist and carry it back to the compressor.

Refrigerating capacity: The rate at which a system can remove heat. Usually stated in tons or British thermal units per hour.

Refrigerating effect: The amount of heat a given quantity of refrigerant will absorb in changing from a liquid to a gas at a given evaporating pressure.

Refrigeration: In general, the process of removing heat from an enclosed space and maintaining that space at a temperature lower than its surroundings.

Refrigeration cycle: The complete operation involved in providing refrigeration.

Reheat: To heat air after dehumidification if the temperature is too low.

Relief valve: A valve designed to open at excessively high pressures to allow the refrigerant to escape safely.

Remote bulb: A part of the expansion valve or control thermostat that assumes the temperature of the suction gas at the point where the bulb is secured to the suction line. Any change in the suction gas superheat at the point of bulb application tends to operate the valve or thermostat in a compensating direction.

Remote-bulb thermostat: A controller that is sensitive to changes in temperature.

Remote power element control: A control whose sensing element is located separate from the mechanism it controls.

Remote system: A refrigeration system whose condensing unit is located away from the conditioned space.

Restrictor: A reduced cross-sectional area of pipe that produces resistance or a pressure drop in a refrigeration system.

Retrofit: Term used in describing reworking an older installation to update it with modern equipment or to meet new code requirements.

Return air: Air taken from the conditioned space and brought back to the conditioning equipment.

Reverse-cycle defrost: A method of reversing the flow of refrigerant through an evaporator for defrosting purposes.

Reversing valve: Device used to reverse direction of the flow, depending on whether heating or cooling is desired.

Riser: A vertical tube or pipe that carries refrigerant in any form from a lower to a higher level.

Riser valve: A valve used for the manual control of the flow of refrigerant in a vertical pipe.

Rotary blade (VANES) compressor: Mechanism for pumping by revolving blades inside cylindrical housing.

Rotary compressor: Mechanism that pumps fluid by using rotary motion.

Rotor: The rotating or turning part of an electric motor or generator.

Saddle valve: A valve equipped with a body that can be connected to a metal refrigerant line. The valve body is shaped so it may be brazed or clamped onto refrigerant tubing.

Safety can: Approved container of not more than 5-gallon (18.9-liter) capacity that has a spring-closing lid and spout cover. It is designed to relieve internal pressure safely when exposed to fire.

Safety control: Any device that allows the refrigeration or air conditioning unit to stop when unsafe pressures or temperatures exist.

Safety factor: The ratio of extra strength or capacity to the calculated requirements to ensure freedom from breakdown and ample capacity.

Safety plug: A device that releases the contents of a container to prevent rupturing when unsafe pressures or temperatures exist.

Safety valve: A quick-opening safety valve used for the fast relief of excessive pressure in a container.

Saturation: A condition existing when a substance contains the maximum amount of another substance that it can hold at that particular pressure and temperature.

Scale: Homogeneous crystalline deposits formed on surfaces in contact with water.

Scavenger pump: Mechanism used to remove fluid from a container.

Schrader valve: A spring-loaded valve that permits fluid to flow in only one direction when the center pin is depressed.

Scotch yoke: A mechanism that changes rotary motion into reciprocating motion used to connect a crankshaft to a piston in some types of compressors.

Screw pump: A compressor that is constructed of two mating revolving screws.

Sealed unit: A motor compressor assembly in which the motor and compressor operate inside a common sealed housing.

Seal, front: The part of refrigeration valve that forms a seal with the valve button when the valve is in the closed position.

Seal leak: The escape of oil and/or refrigerant at the junction of a shaft or housing.

Seal, shaft: A device used to prevent leakage between shaft and case.

Seasonal energy efficiency ratio (SEER): The total cooling of a central unitary air conditioner or unitary heat pump in British thermal units during its normal annual usage period for cooling divided by the total electric energy input in watt-hours during the same period.

Seat: The portion of a valve mechanism against which the valve stem rests to effect shutoff.

Secondary refrigerating system: Refrigerating system in condenser is cooled by evaporator of another or primary refrigerant system.

Second law of thermodynamics: A law that states that heat will flow from an object at a higher temperature to an object at a lower temperature.

Seebeck effect: When two different adjacent metals are heated, an electric current is generated between the metals.

SEER: The abbreviation for seasonal energy efficiency ratio.

Selective absorber surface: Surface used to increase temperature of a solar collector.

Self-contained air conditioning unit: An air conditioner containing a condensing unit, evaporator, fan assembly, and complete set of operating controls within its casing.

Semihermetic compressor: A hermetic compressor on which minor field service operations can be performed.

Sensible heat: Heat that causes a change in temperature of a substance but not a change of state.

Sensible heat ratio: The percentage of total heat removed that is sensible heat. It is usually expressed as a decimal and is the quotient of sensible heat removed divided by total heat removed.

Sensor: A device that undergoes a physical change or a characteristic change as surrounding conditions change.

Separator: A device to separate one substance from another.

Separator, oil: A device to separate refrigerant oil from refrigerant.

Glossary

Sequence controls: A group of controls that act in a series or in a timed order.

Serpentine: The arrangement of tubes in a coil to provide circuits of the desired length. Intended to keep pressure drop and velocity of the substance passing through the tubes within the desired limits.

Serviceable hermetic: Hermetic unit housing containing motor and refrigerant compressor assembly by use of bolts or cap screws.

Service valve: Manually operated valve mounted on refrigerating systems used for service operation.

Servo: A servomechanism in a low-power device, either electrical, hydraulic, or pneumatic, that puts into operation and controls a more complex or more powerful system.

Set point: The temperature to which a thermostat is set to result in a desired temperature.

Shell and coil: A designation for heat exchangers, containers, and chillers consisting of a tube coil within a shell or housing.

Shell and tube: A designation for heat exchangers, containers, and chillers consisting of a tube bundle within a shell or casing.

Shell-and-tube flooded evaporators: These evaporators use water flow through tubes built into cylindrical evaporator, or vice versa.

Shell-type condenser: Cylinder or receiver that contains condensing water coils or tubes.

Short cycling: Refrigerating system that starts and stops more frequently than it should.

Shroud: Housing over condenser, evaporator, or fan.

Sight glass: A glass tube or window in refrigerating mechanism.

Significant New Alternatives Policy (SNAP): A policy to identify alternatives to ozone-depleting substances (ODS) and then publish a list of acceptable and unacceptable substances.

Single-stage compressor: Compressor having only one compressive step between inlet and outlet.

SI unit system (Le Système International d'Unites): Metric system of measurement adopted by most technical industries throughout the world.

Sling psychrometer: A device with a dry-bulb and a wet-bulb thermometer that is moved rapidly through the air to measure humidity.

Slug: (1) Unit of mass equal to the weight (in U.S. units) of object divided by 32.2 (acceleration due to the force of gravity). (2) Detached mass of liquid or oil, which causes an impact or hammering noise in a circulating system.

Slugging: A condition in which a quantity of liquid enters the compressor cylinder, causing a hammering noise.

Small appliance: Defined by the Environmental Protection Agency (EPA) as any of the following products that are fully manufactured, charged, and hermetically sealed in a factory with five pounds or less of refrigerant: refrigerators and freezers designed for home use, room air conditioners (including window air conditioners and packaged terminal air conditioners), packaged terminal heat pumps, dehumidifiers, under-the-counter ice makers, vending machines, and drinking water coolers.

SNAP: Abbreviation for Significant New Alternatives Policy.

Solar collector: Device used to trap solar radiation, usually using an insulated black surface.

Solar energy systems: Systems used to collect, convert, and distribute solar energy in forms useful within a business or residence.

Solar heat: Heat created by visible and invisible energy waves from the sun.

Solder: To join two metals by the adhesion process using a melting temperature less than 800°F (427°C).

Glossary

Soldering: Joining two metals by adhesion of a metal with a melting temperature of less than 800°F (427°C).

Solenoid valve: Electromagnet with a moving core that serves as a valve or operates a valve.

Solution: (1) A homgeneous mixture of two or more substances which may be solids, liquids, gases, or a combination of these. (2) The state of being dissolved.

Sone: Calculated sound loudness rating.

Sound tracer: Instrument that helps locate sources of sound.

Specific gravity: The weight of a liquid compared with water, which is assigned the value of 1.0.

Specific heat: Ratio of quantity of heat required to raise the temperature of a body to that required to raise the temperature of equal mass of water.

Specific volume: Volume per unit mass of a substance. The volume per unit of mass. Usually expressed as cubic feet per pound.

Splash system, lubrication: The method of lubricating moving parts by agitating or splashing the oil around in the crankcase.

Splash system, oiling: Method of lubricating moving parts by agitating or splashing oil in the crankcase.

Split system: A refrigeration, air-conditioning, or heat pump installation that places the condensing unit outside or away from the evaporator. An air-conditioning or heat pump system that is split into two sections; an outdoor section and an indoor section.

Spray cooling: Method of refrigerating by spraying expendable refrigerant or by spraying refrigerated water.

Squirrel cage: A centrifugal fan that has blades parallel to fan axis and moves air at right angles or perpendicular to fan axis.

Standard air: Air whose temperature is 68°F (20°C), with a relative humidity of 36 percent, and a pressure of 14.7 psig (101.36 kPa).

Standard atmosphere: A condition existing when the air is at 14.7 psia (101.36 kPa) pressure at 68°F (20°C) temperature, and 36 percent relative humidity.

Standard conditions: Conditions used as a basis for air conditioning calculations: temperature of 68°F (20°C), pressure of 29.92 inches of mercury (Hg), and a relative humidity of 30 percent.

Static head: The pressure due to the weight of a fluid in a vertical column or, more generally, the resistance due to lift.

Stationary blade compressor: A rotary pump that uses a nonrotating blade inside pump to separate intake chamber from exhaust chamber.

Steam: Water heated to the vaporization point.

Steam jet refrigeration: A refrigeration system that uses a steam venturi to create low pressure on a water container, causing water to evaporate at low temperature.

Steam trap: An automatic valve that traps air but allows condensate to pass while preventing passage of steam.

Stethoscope: An instrument used to detect sounds and locate their origin.

Strainer: A device, such as a screen or filter, used to retain solid particles while passing liquid.

Stratification of air: The condition in which there is little or no air movement in a room.

Subcooling: The cooling of liquid refrigerant below its condensing temperature.

Subcooling coils: A supplementary coil in an evaporative condenser, usually a coil or loop immersed in the spray water tank, that reduces the temperature of the liquid leaving the condenser.

Glossary

Sublimation: A condition in which a substance changes from a solid to a gas without passing through the liquid state.

Substance: Any form of matter or material.

Suction line: The tube or pipe that carries the refrigerant vapor from the evaporator to the compressor.

Suction pressure control valve: A device, located in the suction line, that maintains constant pressure in the evaporator.

Suction service valve: A two-way manually operated valve located at the inlet to the compressor used to control suction gas flow and to service the unit.

Suction side: The low-pressure side of the system extending from the refrigerant control, through the evaporator, to the inlet of the compressor.

Superheat: (1) The temperature of vapor above its boiling point as a liquid at that pressure. (2) The difference between the temperature at the evaporator outlet and the temperature of the refrigerant in the evaporator.

Superheater: A heat exchanger used to remove heat from liquid refrigerant going to the evaporator and adding it to superheat the vapor leaving the evaporator.

Technician: Defined by the Environmental Protection Agency (EPA) as (1) any person who performs maintenance, service, or repair that could reasonably be expected to release class I (CFC) or class II (HCFC) substances from appliances, except for MVACS, into the atmosphere; (2) any person performing disposal of appliances, except for small appliances, MVACS, and MVAC-like appliances, that could be reasonably expected to release class I or II refrigerants from appliances into the atmosphere.

Temperature sensing bulb: A bulb containing a volatile fluid and bellows or diaphragm.

Therm: A quantity of heat that is equal to 100,000 Btu.

Thermal relay: An electrical control used to open or close a refrigeration system electrical circuit.

Thermocouple: A device that generates electricity, using the principle that if two dissimilar metals are bonded together and heated a voltage will develop across the open ends.

Thermocouple thermometer: An electrical instrument using a thermocouple as a source of electrical flow for determining temperature.

Thermodisk defrost control: An electrical switch with a bimetallic disk controlled by temperature changes.

Thermodynamics: The part of science that deals with the relationships between heat and mechanical action.

Thermoelectric refrigeration: A refrigerator mechanism that depends on the Pelter effect.

Thermometer: A device for measuring temperatures.

Thermostat: A temperature-sensitive switch that is used in the control of heating or cooling.

Thermostatic control: A device that operates a system or a part of a system based on temperature change.

Thermostatic expansion valve (TXV): A metering device controlled by temperature and pressure within the evaporator.

Thermostatic motor control: A device used to control the cycling of a unit based on temperature changes.

Thermostatic water valve: A valve used to control the flow of water through a system, actuated by temperature difference.

Three-way valve: A multiorifice flow control with three fluid flow openings.

Throttling: The expansion of gas through an orifice or controlled opening without the gas performing any work as it expands.

Glossary

Timers: A clock-operated mechanism used to control opening and closing of an electrical circuit.

Timer-thermostat: A thermostat control that includes a clock mechanism to automatically control temperature changes based on the time of day.

Ton: A term often used for the capacity of an air-conditioning system equal to 12,000 Btu per hour (Btu/h).

Ton of refrigeration: The refrigerating effect equal to the melting of one ton of ice in 24 hours: 288,000 per day, 12,000 Btu per hour, or 200 Btu per minute.

Total heat: The sum of both the sensible and latent heat.

Transmission: The loss or gain from a building through exterior components such as windows, walls, and floors.

Trichlorotrifluoromethane: Chemical name of refrigerant R-113.

Triple point: The pretemperature condition in which a substance is in equilibrium in solid, liquid, and vapor states.

Troposphere: The part of the atmosphere immediately above Earth's surface in which most weather disturbances occur.

Tube, constricted: Tubing reduced in diameter.

Tube-within-a-tube: A water-cooled heat exchanger in which a tube is placed inside a larger tube; refrigerant is passed through the outer tube and water through the inner tube.

Tubing: Fluid-carrying pipe that has a thin wall.

Two-temperature valve: A pressure-operated valve used in the suction line on multiple refrigerator installations that maintains evaporators in system at different temperatures.

Two-way valve: A valve with one inlet port and one outlet port.

Unacceptable: A Significant New Alternatives Policy (SNAP) designation meaning that it is illegal to use a product as a substitute for an ozone-depleting substance (ODS) in a specific end use. For example, HCFC-141b is an unacceptable substitute for CFC-11 in building chillers. Note that all SNAP determinations apply to the use of a specific product as a substitute for a specific ODS in a specific end use.

Use restriction: A Significant New Alternatives Policy (SNAP) term that includes both use conditions and narrowed use limits.

Ultraviolet (UV): Invisible radiation waves with frequencies shorter than wavelengths of visible light and longer than X rays.

Unitary system: A heating/cooling system factory assembled in one package and usually designed for conditioning one space or room.

Urethane foam: Type of insulation that is foamed in between inner and outer walls of a container.

Vacuum: Pressure lower than atmospheric pressure.

Vacuum activators: Dampers and control valves used in automotive air conditioning systems; controlled by the vacuum created by engine intake manifold vacuum.

Vacuum pump: A special high-efficiency device used for creating high vacuums for testing or drying purposes.

Valve: A device used for controlling fluid flow.

Valve, expansion: A type of refrigerant metering that maintains constant pressure in the low side of refrigerating mechanism.

Valve plate: The part of a compressor located between the top of the compressor body and the head containing suction and discharge valves.

Valve, service: A mechanical device used to access the system to check pressures, service, and charge refrigerating systems.

Valve, solenoid: A valve made to work by magnetic action through an electrically energized coil.

Glossary

Valve, suction: A valve in the compressor that allows vaporized refrigerant to enter cylinder and prevents its return.

Valve, water: A valve that regulates the flow of water to cool the system while it is running.

Vapor: (1) Vaporized refrigerant, also called gas. (2) A gas that is often found in its liquid state while in use.

Vapor barrier: A thin plastic or metal foil used in air-conditioned structures to prevent water vapor from penetrating insulating material.

Vaporization: A change of liquid into a gaseous state.

Vapor lock: A condition in which liquid is trapped in a line because of a bend or improper installation, preventing liquid flow.

Vapor, saturated: The vapor condition that will result in condensation into droplets of liquid if its temperature is reduced.

V-belt: A type of belt used in refrigeration work having a contact surface with a pulley that is in the shape of a V.

Velocimeter: An instrument that measures air speeds using a direct-reading air-speed indicating scale.

Velocity: The quickness or rapidity of motion, swiftness, speed.

Ventilation: Forced airflow between one area and another.

Vibration arresters: Soft or flexible substances or devices that will reduce the transmission of a vibration.

Viscosity: The measurement of thickness of oil or its resistance to flow.

Voltmeter: An instrument for measuring voltage in electrical circuits.

Volumetric efficiency: A term used to express the relationship between the actual performance of a compressor or pump and the calculated performance based on its displacement.

Vortex tube: A mechanism for accomplishing a cooling effect by releasing compressed air through a specially designed tube.

Walk-in cooler: A commercial refrigerated space kept below room temperature.

Water-cooled condenser: A heat exchanger that is cooled through the use of water flow.

Water defrosting: The use of water to melt ice and frost from an evaporator during the off-cycle.

Water hammer: A noise generated by water back pressure when a valve is closed.

Wax: An ingredient in many lubricating oils that may separate from the oil if sufficiently cooled.

Wet bulb: A device used in the measurement of relative humidity.

Wet-bulb temperature: A measure of the amount of moisture for an air sample.

Wobble plate–swash plate: Type of compressor designed to compress gas, with piston motion parallel to crankshaft.

Zero ice: Trade name for dry ice.

Zone controls: Controls used to maintain each specific area or zone within a building at a desired condition.

SECTION 6: ANSWER KEYS

CORE PRACTICE EXAM ANSWER KEY

1.	B	20.	C	39.	B	58.	B
2.	B	21.	C	40.	C	59.	C
3.	D	22.	A	41.	D	60.	B
4.	A	23.	C	42.	B	61.	D
5.	B	24.	A	43.	A	62.	B
6.	C	25.	A	44.	A	63.	A
7.	B	26.	C	45.	D	64.	A
8.	C	27.	A	46.	B	65.	C
9.	D	28.	D	47.	A	66.	B
10.	C	29.	A	48.	D	67.	C
11.	A	30.	D	49.	B	68.	B
12.	D	31.	D	50.	A	69.	B
13.	D	32.	C	51.	D	70.	C
14.	D	33.	A	52.	A	71.	D
15.	C	34.	D	53.	B	72.	B
16.	A	35.	A	54.	C	73.	D
17.	D	36.	C	55.	B	74.	D
18.	D	37.	C	56.	B	75.	C
19.	B	38.	C	57.	A		

TYPE I PRACTICE EXAM ANSWER KEY

1.	D	20.	D	39.	D	58.	A
2.	B	21.	C	40.	C	59.	C
3.	B	22.	D	41.	A	60.	D
4.	D	23.	B	42.	A	61.	A
5.	D	24.	B	43.	D	62.	D
6.	C	25.	D	44.	A	63.	C
7.	D	26.	C	45.	A	64.	A
8.	A	27.	D	46.	B	65.	B
9.	C	28.	B	47.	C	66.	C
10.	D	29.	B	48.	B	67.	A
11.	D	30.	D	49.	D	68.	D
12.	C	31.	C	50.	C	69.	B
13.	D	32.	C	51.	D	70.	A
14.	B	33.	B	52.	D	71.	D
15.	A	34.	B	53.	A	72.	C
16.	C	35.	A	54.	C	73.	B
17.	A	36.	B	55.	D	74.	A
18.	A	37.	A	56.	B	75.	A
19.	B	38.	D	57.	C		

TYPE II PRACTICE EXAM ANSWER KEY

1.	B	20.	C	39.	A	58.	C
2.	A	21.	D	40.	D	59.	B
3.	A	22.	D	41.	A	60.	B
4.	C	23.	C	42.	C	61.	D
5.	A	24.	D	43.	B	62.	B
6.	D	25.	B	44.	A	63.	C
7.	A	26.	C	45.	D	64.	D
8.	A	27.	D	46.	B	65.	D
9.	A	28.	C	47.	A	66.	D
10.	C	29.	B	48.	C	67.	A
11.	C	30.	D	49.	D	68.	B
12.	C	31.	A	50.	D	69.	C
13.	D	32.	A	51.	C	70.	B
14.	B	33.	B	52.	B	71.	A
15.	D	34.	B	53.	A	72.	D
16.	A	35.	A	54.	A	73.	D
17.	B	36.	C	55.	C	74.	D
18.	C	37.	B	56.	B	75.	A
19.	C	38.	D	57.	A		

TYPE III PRACTICE EXAM ANSWER KEY

1.	C	20.	B	39.	C	58.	C
2.	D	21.	D	40.	D	59.	D
3.	C	22.	C	41.	B	60.	B
4.	A	23.	D	42.	C	61.	C
5.	C	24.	C	43.	A	62.	D
6.	A	25.	B	44.	C	63.	C
7.	D	26.	A	45.	A	64.	B
8.	A	27.	B	46.	D	65.	C
9.	D	28.	D	47.	A	66.	A
10.	B	29.	A	48.	C	67.	B
11.	B	30.	C	49.	D	68.	A
12.	D	31.	B	50.	B	69.	D
13.	B	32.	A	51.	B	70.	B
14.	B	33.	D	52.	A	71.	B
15.	D	34.	A	53.	C	72.	C
16.	D	35.	B	54.	B	73.	C
17.	C	36.	A	55.	B	74.	D
18.	A	37.	B	56.	D	75.	A
19.	A	38.	C	57.	B		